William Lawrence Woodruff

Climatography of the Salt River Valley Region of Arizona

William Lawrence Woodruff

Climatography of the Salt River Valley Region of Arizona

ISBN/EAN: 9783742810281

Manufactured in Europe, USA, Canada, Australia, Japa

Cover: Foto ©Klaus-Uwe Gerhardt /pixelio.de

Manufactured and distributed by brebook publishing software
(www.brebook.com)

William Lawrence Woodruff

Climatography of the Salt River Valley Region of Arizona

CLIMATOGRAPHY

OF THE

SALT RIVER VALLEY REGION OF ARIZONA

THE LAND OF HEALTH AND SUNSHINE

STUDIES FOR PHYSICIANS AND LAYMEN, WITH METEOROLOGICAL DATA COMPILED FROM THE
REPORTS OF THE UNITED STATES WEATHER BUREAU IN TABULAR FORM, COMPARING
THIS WITH OTHER PARTS OF THE UNITED STATES, VITAL STATISTICS, AND LIST
OF DISEASES EITHER BENEFITED OR CURED IN THIS SALUBRIOUS
CLIMATE, AND COMPLETE AND ACCURATE DATA FOR THE
INFORMATION OF INVALID OR HOME SEEKER.

BY

WM. LAWRENCE WOODRUFF, M. D.
PHOENIX, ARIZONA

CHICAGO
R. R. DONNELLEY & SONS COMPANY
1898

THE CLIMATE OF PHOENIX AND THE SALT RIVER REGION OF ARIZONA.

An article by Wm. Lawrence Woodruff, M.D., of Phoenix, Arizona, printed in the Hahnemannian Monthly for December, 1895, reprinted in the Scientific American (Supplement) of January, 1896, and reprinted in the Sanitarian for May, 1896, and reprinted in The Arizonian for January, 1896.

The inquiries about Phoenix and the Salt River Valley as a health resort are becoming so numerous that I take it the profession at large will welcome *facts* concerning this valley, and facts only I will endeavor to state in this article. My aim is to cover the ground fully with the most reliable data attainable.

Phoenix and the Salt River Valley are situated in latitude 33° north, in the southwest quarter of Arizona. The valley is from five to seventy-five miles wide, and about two hundred miles long, and throughout its entire length and breadth has a climate claimed to be the best in the world. To rightly appreciate the claims of this valley as a health resort, we must for a moment look at the physical geography of this region. There are high mountain ranges to the north and east, also the Sierra Nevada and Coast Ranges to the west, with a short spur of low mountains to the south. The high mountain ranges protect this section from all cold winds, and to this protection from cold, nature has added yet another feature, which is mainly the cause of the phenomenal climatic conditions found in this

region, namely, proximity to the Gulf of California. The
Salt River Valley, with the Gila Valley, its extension to
the southwest, is an open valley with continuous mountain
chains of more or less altitude on either side, and practically maintains these characteristics clear to the head of the
gulf. The Gulf of California, with the Coast Range on its
west to protect it from cold, northwest winds, and a lower
mountain range east of it is so situated that it catches and
retains the warm winds and ocean currents from the Indian
Ocean and the equatorial Pacific, and passes them up to
the head of the gulf, and, consequently, is largely responsi-
ble for the warm, mild winters. It will be seen by the
above how nature has provided a channel whereby, in this
southwest corner of the United States, she has reproduced
a climate tropical in all its essential parts, with none of the
drawbacks of the tropics, namely, excessive humidity and
malaria. Here, right in our midst, nature has given a
climate as mild and balmy as that of the tropical Pacific
islands, and with the same even temperature, and at the
same time at an altitude of only eleven hundred feet, a
dryness of atmosphere equaled by few localities and ex-
celled by none other in the civilized world. It will now be
understood how a climate that seldom gives a temperature
at the freezing point, with rarely a cloudy day—there is
less than one in ten during the winter, and for weeks at a
time during the summer there is not a cloud in the sky—is
possible at this latitude. Here is found every element that
goes to make up a perfect climate. The best proof on this
point is the exceptionally low death-rate, which is 8 11-100
per 1,000 inhabitants. This sun-kissed valley has but two
seasons—the winter season, which is a happy blending of

fall into spring, and the summer, which commences about
May 1st, and continues until about October 1st. The summer days are bright, clear and hot, with a maximum daily
temperature ranging from 96° to 112°. It is as rare for
the mercury to go above this in summer as it is rare for it
to go below the freezing point in winter. There is usually
some little rain in the latter part of July or during August,
usually in showers, possibly averaging an inch of rainfall
during the summer season. To rightly appreciate the
effects of the summer heat, one must recognize the difference between a wet and a dry-bulb thermometer. The difference is usually from 20° to 30°. Here, the reading of
the wet-bulb gives our actual sensible heat, while in more
humid countries the reading of the dry-bulb is so nearly
like that of the wet-bulb that the difference is rarely perceptible. The average humidity is only about 30 per cent.
for the year, and there are weeks at a time during the summer when it will run far below this point. This is the reason, coupled with the fact that there is always a gentle
breeze stirring, why our summers are not only endurable,
but, in fact, do not cause as much discomfort or prostration as is experienced in other parts of the country. The
summer months are the healthiest of the year. During
these months the death rate is only one-third of one per
cent. Bowel troubles and fevers are almost unknown during the heated term, there being less than two deaths per
month from all forms of bowel troubles among infants in a
population of 14,000. Is there another place in the world
that can make such a showing? During these months perspiration is very copious, and, owing to the very dry air,
evaporation is instantaneous and a material aid to comfort.

With this statement the fact will be readily understood that rheumatism, kidney diseases, and diseases of the respiratory tract make their greatest improvement during this half of the year. This is especially so with persons suffering from insomnia and nervous prostration. Sunstroke is unknown, and it is as safe for people to come here during the heated term as at any other time of the year.

Now, as to the winter months. The visitor will find the days balmy, dreamy, restful; the air pure, dry, bracing; the nights cool and delightful. Save during the rainy seasons, it is perfectly safe and comfortable to be out of doors day and night. The rainy season usually lasts a week or so, and the rainfall is not heavy. The annual precipitation is something less than seven inches. The following table shows the maximum temperature for a period from December 31, 1894, to January 9, 1895, inclusive, at several places. An examination of this table will show that

Date.	Phoenix, Ariz.	Los Angeles, Cal.	Jacksonville, Fla.	Tampa, Fla.	St. Augustine, Fla.	Hot Springs, Ark.	Nice, France.	Malta.	Cairo, Egypt.	St. Moritz, Switzerland.	Rome, Italy.
December 31, 1894	74	61	44	60	28	39	51	59	65	50	44
January 1, 1895	74	63	49	60	28	49	53	59	65	50	39
" 2, "	68	61	60	58	52	49	53	59	65	48	35
" 3, "	72	61	60	68	60	50	46	52	65	57	35
" 4, "	70	55	61	52	50	50	46	57	63	52	35
" 5, "	68	55	54	62	52	49	44	54	64	52	28
" 6, "	70	53	62	62	52	49	46	54	63	57	--
" 7, "	68	57	75	62	52	49	42	54	65	50	41
" 8, "	65	62	74	62	68	48	41	59	64	42	39
" 9, "	64	68	68	78	63	49	50	54	64	42	39
Range of Temp. for the 10 days	10	15	31	26	40	11	12	7	2	15	16

Phoenix has the most even temperature of all the places named, with but two exceptions, one being Cairo, Egypt, whose highest temperature is 65°—but one degree above our lowest, 64°—and Malta, with 59° as the highest point reached, being 5° below our lowest point. These two places—as, indeed, do all the rest named—have a damp, moist atmosphere, which greatly increases the perceptible difference in the range of temperature.

This valley has everything that goes to make up a perfect winter home. It has the minimum of rainfall — 7 inches per annum; second, the minimum of atmospheric moisture—30 per cent. humidity; third, it has the minimum air movement—its annual average is less than 2½ miles per hour, and is generally from the southwest; fourth, the minimum of death-rate, being but 8 11-100 per 1,000 inhabitants; fifth, the minimum of malaria, there being none; sixth, low altitude—1,100 feet above the sea-level; seventh, the maximum of sunshine—an average of nine days out of ten of bright sunshine, when out-of-door life is enjoyable and healthful. We have here within easy reach, and within the bounds of our own country, all the merits ascribed to Italy or Egypt, with none of their drawbacks. We have all that Florida enjoys, with none of her moist, sticky atmosphere and none of her malaria. We have the same balmy air and even temperature of California, without her fogs, dampness, or malaria. We have the same dry, bracing air that has Colorado, without her blizzards and high altitudes. We have all, and infinitely more, of all the good things claimed for these localities, without their unfavorable conditions. There may be a few localities where the actual difference in temperature between day and night is less than in the Salt River Valley, but these places have much greater humidity. As in summer, so here in winter, with our very dry air, the perceptible difference between day and night temperatures, and the actual discomfort experienced thereby, is much less than is the case in localities with more moisture in the air. Situated in the midst of this valley, about 150 miles from the head of the Gulf of California, 1,100 feet above the sea-level, lies Phoenix, the capital of Arizona and the metropolis of the Salt River Valley. It is the healthiest city in the known world, and is surrounded by a prosperous and constantly growing farming community. It has all the modern improvements and the snap and vim of the young metropolis. Her citizens are quiet, peaceable and law-abiding, and ready to receive with true hospitality those who seek her perpetual sunshine. The town is making a phenomenal growth, in spite of the hard times, and will soon have the best of accommodations for the health-seeker, who will find the pure, dry, warm, health-giving air free for all.

The following comparative mortality table shows the yearly deaths in 1,000 inhabitants in the cities named. It will be noticed that Phoenix stands at the head of the list. Phoenix, Ariz., 8 11-100; Los Angeles, Cal., 14 40-100; Long Branch, N. J., 9 88-100; Atlantic City, N. J., 18 38-100; St. Paul, Minn., 9 60-100; Minneapolis, Minn., 9 40-100; San Bernardino, Cal., 11 30-100. There are no public records from which an accurate table of vital statistics can be compiled. The records of the undertakers in the territory named are accurate and complete for the past three years, and include, with very few exceptions, all the deaths in that territory during the period covered. These records have been kindly placed at my disposal, and from them I have prepared a table with a great deal of care. For all practical purposes it is accurate and reliable.

VITAL STATISTICS OF THAT PART OF THE SALT RIVER VALLEY NORTH OF THE SALT RIVER, WEST OF THE VERDE RIVER AND EAST OF THE AGUA FRIA RIVER, COVERING A TERRITORY OF 250 SQUARE MILES, AND INCLUDING THE CITY OF PHOENIX. THE POPULATION ON A CONSERVATIVE BASIS IS PUT AT 14,000; FOR 1895, AT 15,000; FOR 1896, 16,000.

	1892	1893	1894	1895	1896
Total number of deaths	133	185	168	141	205
Transients	29	38	41	47	78
Accidental deaths	10	15	7	13	15
Among residents	94	132	120	81	112
Percentages, fractions 1%	3/4	4/9	6/7	1/8	3/8
CLASSIFIED BY AGES.					
Deaths under 5 years of age	28	59	33	29	38
Deaths over 70 years of age	12	8	13	7	10
Deaths over 50 years of age	31	32	36	19	43
DURING THE SUMMER MOS.— JUNE – SEPTEMBER.					
Total	41	75	54	58	75
Transients and accidentals	8	21	13	23	25
Residents, from natural cause	23	54	41	35	50
Percentages, fractions 1%	1/3	2/8	1/3	1/3	1/8
Under 5 years of age	6	28	13	14	19
Under 5, of bowel trouble	6	11	9	5	8
CAUSES OF DEATH.					
Stomach and bowel disease	10	30	21	14	15
Nervous and brain disease	17	8	4	8	6
Typhoid fever	2	4	4	2	4
Scarlet fever	1	3	0	0	0
Measles	0	4	0	0	0
Diphtheria	0	5	2	0	0
Heart disease	8	1	7	3	8
Disease respiratory organs	50	73	61	56	82
Old age	4	4	6	4	3
All other causes	40	56	58	54	87

NOTE.—Deaths designated as transients are only those of persons who have been here but a brief period prior to their decease, coming here as a last resort in the advanced stages of diseases of the respiratory organs, which accounts for the large number of deaths under this head. A large number of those claimed as residents ought properly to have been included in the transient class.

The following statement will illustrate the general healthfulness of this valley under one set of conditions:

PHOENIX, September 28, 1895.

Dear Doctor:—I have been working large gangs of men in construction-work of different kinds, for the last fourteen years, in the Northwest and in Canada. Last spring I brought in a large gang of men from Minnesota, and for the last six months have been working them with others in your valley, and never, in all my experience, has the percentage of sickness been so low as during these past six months.

(Signed)   S. R. H. ROBINSON, Superintendent,
Minnesota and Arizona Construction Company.

Now, as to diseased conditions: Asthmatics usually receive prompt relief and a permanent cure. The dry, warm air, and low altitude agree with them perfectly. If there is a recurrence it is during the rainy season, and is usually but slight, to disappear again as soon as the usually dry atmospheric conditions prevail. This is equally so of aphonia, bronchitis, and laryngitis; and, in fact, of all diseases of the respiratory organs. Tuberculosis, by the dry, hot air of summer, is checked in its development; and, if the patient is not in the last stages, a continuous residence under these favorable conditions, will greatly prolong life, and often eventually bring about a cure. Let me say here, if the patients have entered the last stage of the disease, in the interest of humanity keep them at home. This cannot be emphasized too strongly. There they will have more comforts; and the radical change of climate, with the long, and tiresome journey necessary in reaching here, only tends to materially hasten the end. During the winter months, this class of patients, in common with all others, may reasonably expect to hold their own, and usually make substantial gains. It will readily be perceived by a careful perusal of this article, that there is greater reason to expect beneficial results in all diseased conditions from a sojourn in this climate, than in any other winter resort. While this is undoubtedly so, it is equally true that the hot, dry air of summer produces the best results. In heart diseases we find the cooler weather of winter the most beneficial. In some cases the reverse is true. The hotter and dryer it gets, the more comfortable the patient becomes. This is especially so where the disease is complicated with diseased kidneys or rheumatic diathesis. Catarrhal conditions of head and throat are most relieved during the summer, especially the moist varieties. Diseases of the digestive tract, dyspepsia, chronic dysentery, and diarrhea, do exceedingly well here, and are usually promptly relieved. This is doubly true during the hot months. The summer conditions, of high temperature and low humidity, cause a determination of blood to the surface, maintaining it there for months at a time, and thereby entirely relieving all internal congestions. Kidney troubles are so prevalent I must not forget to mention, that during the heated term the kidneys excrete less than one-half of the normal quantity of urine. During this period of rest, the unloading of the effete material of the system is carried on by the sweat-glands of the skin, and a healthy equilibrium is maintained. This continuous high temperature and very dry air keeps the blood at the surface, thereby making the sweat-glands very active. Perspira-

tion is constant and copious, and, by its instant evaporation, keeps the surface cool and the bodily temperature at
normal. These conditions are very advantageous to diseased kidneys, giving them a much needed rest, and an
opportunity to recuperate. When to this is added a drink-
ing-water, pure, wholesome, and devoid of all alkali, it is
easily understood why this valley is fast getting an enviable
reputation for the alleviation and cure of all forms of this
disease. In rheumatic affections, while in winter patients are
made very comfortable, it is in summer that the constant
free perspiration maintained for months without ceasing,
entirely eliminates from the system all morbid material.
In diseases of the nervous system, so prevalent in this age,
this climate is a true panacea. This is especially so of
persons suffering from insomnia and nervous prostration.
Here, again, the best results are during the summer
months. The universal verdict is, "I have nowhere else slept
as I do here." This is the universal expression. The tired-
out, starved nerves, over-worked and over-wrought, experience in this balmy air the perfect relaxation and rest they
so long have been in need of. The dry, hot air of summer
seems to quiet the nervous system, is soothing, restful, and
when to this a voracious appetite is added, with perfect di-
gestion, which is the only epidemic during this season, the
results are understood without further elaboration. Fin-
ally, the perfect summer nights soothe and rest one's
nerves as does nothing else in all the world.

THE CLIMATE OF SALT RIVER VALLEY.

A paper read before the American Institute of Homœopathy at Detroit, Michigan, in June 1896, by Wm. L. Woodruff, M.D., Phoenix, Arizona, and printed in the "Transactions" of that year at page 994.

That climate stands at the head of the list of favorable conditions requisite for the successful prevention and arrest of the progress of a large number of diseases I think you will all readily admit.

I think it equally true, that, believing the above, the profession are anxious to learn of the best place for the greatest number. That place I claim, and will try to prove, is the Salt River Valley in the southwestern quarter of Arizona, with Phoenix as its largest center of population.

The essential features of climate necessary to meet the requirements of an ideal health resort, for persons suffering with chronic diseases generally, and especially with diseases of lung and throat, kidneys, rheumatism, and conditions of mal-nutrition are, First, a warmth and geniality which enables the weakened subject to spend the greatest possible amount of time in the pure air with the minimum amount of clothing. Second, a degree of dryness of the atmosphere which will insure rapid and easy elimination from the skin, thus relieving the weakened and diseased mucous

13

surfaces from the full task of elimination, which ordinarily they are expected to perform, but under diseased conditions cannot accomplish, in consequence of which inability there is imperfect elimination, and gradual poisoning of the system from the circulation of blood not fully deprived of its effete material. Third, an equability of climate, which does not suddenly go from great extremes of heat to cold, whose night and day temperatures are not too far separated, and where there is so little dampness in the air that the changes in temperature are but little felt. Fourth, the minimum of wind movement, and that with the least possible contamination from decaying vegetable substances, decomposed animal matter, or poisonous gases of whatever origin. Fifth, such a combination of climate, general healthfulness and commercial, industrial and social advantages that the health seeker may live in comfort and with profit, if he be inclined to employ himself in remunerative occupations within the limits of his strength.

Imagine, if you can, a valley varying from 5 to 75 miles in width and 200 miles in length, with continuous mountain chains on either side, running from northeast to southwest, this valley terminating in a gulf whose surface contains 53,000 square miles, whose opening into the equatorial Pacific Ocean is 250 miles wide, and having the same continuous mountain chains. You can then readily understand how the equatorial trade winds sweeping up the west coast of Mexico enter with the tropical ocean currents the confines surrounding the Gulf of California, and that these winds, after sweeping over these 53,000 square miles of tropical waters, form the prevailing winter winds of this vast valley, and in great measure produce the mild, salubrious winters for which the upper part of this valley, known as the Salt River Valley, is fast becoming famous.

To this add continuous high mountain ranges surrounding us on the west, north, and east, and you have a land- or rather rock-locked valley, from which all the cold winter winds are excluded, and if perchance while the blizzard is sweeping over the rest of the country we should feel the edge of it, it can only reach us by the settling down of the upper air currents, and not by a direct blow.

Here in this favored spot—the sun-kissed Valley of the Salt River—you will find a haven of rest and safety for the invalid that fills all the requirements, and the like of which does not exist in any other portion of the known world.

Winters in the Salt River Valley are mild, salubrious, with rarely a severe frost. Out of door life is possible, customary and enjoyable, and excepting the rainy season, which lasts but a few days, one can sleep out of doors with impunity. The invalid can spend every hour of the twenty-four out of doors, or in a tent, not only without risk but with great benefit. The pure, warm, dry air, is invigorating and life-giving, and is indeed Nature's stimulant and tonic.

The days are warm, delightful, sunshiny. A cloudy day is a curiosity, there being rarely more than two or three during the month. There is, I think, no other place in the civilized world where the cloudy days are so few and the sunshine so continuous and perpetual. The following tables will demonstrate this better than anything that I can say:

COMPARATIVE DATA AT PHOENIX, ARIZ., AUGUST, 1895, TO JUNE, 1896.

DATA.	August.	September.	October.	November.	December.	January.	February.	March.	April.	May.	June.
Mean actual temperature	89	82	72	57	49	54	56	62	64	74	--
Mean sensible temperature	70	64	59	49	41	44	44	48	48	54	--
Lowest temp.	65	47	48	34	23	30	28	34	38	45	--
Highest temp.	110	107	93	83	78	79	82	92	89	110	--
Mean rel. humidity, 5 A.M.	61	54	67	81	76	69	65	56	50	41	--
Mean rel. humidity, 5 P.M.	27	29	39	54	40	40	25	21	15	14	--
Percentage of sunshine	85	89	88	81	88	77	87	75	91	89	--
Monthly rainfall (inches)	0.27	0.10	0.80	0.89	0.09	0.46	0.05	0.39	0.05	trace	--

Trace rainfall = too small to measure.
100 = continuous sunshine.
Station established August, 1895.

ARTHUR S. WHITE, Observer in charge.

COMPARATIVE TABLE OF DRY-BULB MEAN TEMPERATURES.

	1895.			1896.				Elevation above sea level.
	Oct.	Nov.	Dec.	Jan.	Feb.	Mar.	Apr.	
Phoenix, A. T.	72.0	57.0	49.0	54.0	56.0	62.0	64.0	1160 feet
San Diego, Calif.	63.8	58.1	54.3	54.5	57.5	57 6	56.0	93 "
San Antonio, Texas.	65.5	55.4	51.7	52 0	54.2	59.0	68.1	679 "
Santa Fe, N. M.	46.9	32.6	23.8	31.7	31.6	39.8	47.7	6998 "
Denver, Colo.	46.4	35.4	31.7	35.2	36 2	35.5	48.6	5287 "
Los Angeles, Calif.	66.0	60.0	56.0	58.0	60.0	57.0	56.0	330 "

COMPARATIVE TABLE OF WET-BULB MEAN TEMPERATURES.

	1895.			1896.			
	Oct.	Nov.	Dec.	Jan.	Feb.	Mar.	Apr.
Phoenix, A. T.	59.0	49.0	41.0	44.0	44.0	48.0	48.0
San Diego, Calif.	59.2	51.4	46.6	49.7	48.9	52.2	50.2
San Antonio, Texas.	55.2	30.4	44 8	45 3	46.4	53 1	62.5
San'a Fe, N. M.	39.2	24.3	14.8	26.8	25.4	30.6	33.8
Denver, Colo.	36.6	28.2	25 2	28.0	29.2	29.6	38 5
Los Angeles, Calif.	52.0	52.0	48.0	52.0	51.0	53.0	50.0

COMPARATIVE TABLE OF MEAN MAXIMUM TEMPERATURES.

	1895.			1896.			
	Oct.	Nov.	Dec.	Jan.	Feb.	Mar.	Apr.
Phoenix, A. T........	93.0	83.0	78.0	79.0	82.0	92.0	89.0
San Diego, Calif......	70.6	68.7	65.1	64.3	67.7	66.7	63.9
San Antonio, Texas.	80.0	67.5	65.0	63.6	67.6	72.1	79.0
Santa Fe, N. H.	59.8	44.0	37.7	42.8	42.5	51.4	60.4
Denver, Colo.........	80.0	75.0	69.0	67.0	68.0	76.0	80.0
Los Angeles, Calif...	76.0	72.6	69.0	68.0	73.0	70.0	67.0

COMPARATIVE TABLE OF MEAN MINIMUM TEMPERATURES.

	1895.			1896.			
	Oct.	Nov.	Dec.	Jan.	Feb.	Mar.	Apr.
Phoenix, A. T........	48.0	34.0	23.0	30.0	30.0	28.0	38.0
San Diego, Calif......	58.2	50.1	44.8	46.7	47.7	49.6	49.1
San Antonio, Texas.	57.5	48.4	43.7	44 5	43.8	49.7	61.6
Santa Fe, N. M.......	39.3	26.1	17.1	25.0	23.6	29.6	35.3
Denver, Colo.........	21.0	2.0	5.0	0.0	9.0	0.0	9.0
Los Angeles, Calif...	55.0	47.6	44.0	47.0	45.0	47.0	46.0

As to the dryness of the atmosphere, there is but one opinion, I believe, as to its being an essential feature of an ideal climate and health resort. In this particular, I can assure you, we excel. This valley is the dryest place available for the health seeker, if not the dryest place in the world. The following table of relative humidity for the seven months just past conclusively demonstrates this fact.

COMPARATIVE TABLE OF MEAN RELATIVE HUMIDITY.
(The rainfall for Phoenix during these seven months is 2.70.)

	1895.			1896.			
	Oct.	Nov.	Dec.	Jan.	Feb.	Mar.	Apr.
Phoenix, A. T........	53.0	68.0	58.0	54.0	45.0	38.0	32.0
San Diego, Calif......	78.2	68.0	56.6	72.6	58.0	71.0	67.0
San Antonio, Texas.	57.2	72.3	56.4	66.5	57.8	69.3	75.1
Santa Fe, N. M.......	51.8	60.0	51.6	51.0	43.1	34.9	19.4
Denver, Colo.........	53 4	44.9	41.4	44.5	46.2	59.7	46.8
Los Angeles, Calif...	82.0	60.0	57.0	68.0	52.0	70.0	67.0

I think a study of this table with the one preceding will disclose conditions extremely favorable to the elimination of effete material from the skin, thus relieving the weakened and overburdened mucous membrane, and internal organs, and thereby favoring recuperative processes. This is especially so during the summer months, to which we will refer later.

Our one weak point is the difference between night and day temperatures. This difference is quite marked, but is much more so measured by the dry-bulb thermometer than by the wet-bulb. The extreme dryness of the atmosphere makes the lower temperature less perceptible than in more moist climates, though there the extremes be considerably less. Owing to the dryness of the air the mid-day temperatures do not seem nearly so high as they actually are, neither do the lower temperatures of night produce the chill one would expect, from looking at the reading of the dry-bulb thermometer.

The actual discomfort from this wide range of temperature is but slight, and its danger largely imaginary. Neither danger or discomfort from this cause is equal to that in a moist climate with a range of temperature not more than one-third as great.

This difference is much less, and indeed exists but in a very small degree, in the higher lands of the foothills and upper sides of the valley. The altitude at Phoenix is 1,100 feet, and in the foothills on the sides of the valley it will run from 300 to 500 feet higher.

The wind movement in the Salt River Valley is so slight as scarcely to be a factor. Our average annual wind movement is but two and 84-100 miles per hour. A wind of twenty-five miles an hour is unknown. The gentlest of zephyrs usually prevail. As on all sides there is but barren mountain and desert, as nothing grows except by irrigation, and as the water is under the perfect control of man, there is no danger from decomposed vegetable matter.

The atmosphere is so dry and pure that animal matter dries up instead of decaying. There being no marshes or stagnant pools there is absolutely nothing but pure, uncontaminated air to breathe.

Now the very best proof that what I have claimed in the above is true, is the low death rate for the valley for the past four years, as shown by the table to be found on page 9 of this book.

Now a word as to the summers in this valley. Accurate data I cannot give you as to temperature, humidity, etc. The Weather Bureau station was only established at Phoenix last fall. This I can say from personal observations extending over four summers, and as corroborated by the said table of vital statistics—that there is not a more healthy place on earth than this same Salt River Valley in the summer time. While about one-third of all the deaths in the United States during the summer months are from bowel troubles among infants, here such deaths average less than two each month in a population of 15,000. Our death rate last summer for the whole five hot months was but one-fourth of one per cent., while the average for the whole country was about 2.2 per cent. Is there any other place that can make such a showing?

To understand our unparalleled healthfulness during the period when the rest of the world is suffering from heat prostration and allied diseases we must for a few moments turn our attention to the study of the difference in the reading of the wet and dry-bulb thermometer.

The better to do this I will quote freely from a recent article by Captain William A. Glassford, Signal Corps, U. S. A., Denver, Colo.:

Every person who has resided in the humid and in the sunshine region knows that there is something wrong with the indications of the thermometer; that there is a marked failure to express, in terms of degrees of temperature, the way in which recorded temperature affects his comfort in the two regions. If the traveler from the East happens to be in Albuquerque, Denver, Salt Lake City, Boise City, or Sacramento, when the thermometer is at or near the 100° point, he must be shown the instrument to be satisfied it is so high, because the discomfort that he is familiar with as a concomitant of such recorded heat in his section is entirely absent. Seeking the cause of this fact he is told that it is accounted for by the absence of humidity. To most people the real reason is still more or less obscure. That 100 degrees makes the man hotter in one place than in the other is accepted as well known; but the amount of this difference in degrees is not at all generally apprehended.

On a nearly north and south line near Wilmington, N. C., and Pittsburg, Pa., the compass bearing is due north; while throughout the arid region it swings from 10° to 20° out of true, due to magnetic variation. What would be thought of the practical experience and science of a surveyor from the Eastern States who, on coming to the arid region, would expect to use a compass reading without knowing or using this magnetic variation? None the less unscientific, if I may not say absurd, when considering the

sensible climatic influence on the human body, is the placing side by side of the recorded thermometric observations
of an arid with a humid region, without applying a correc-
tion or variation factor for dryness and humidity, as is
necessary for the magnetic variation when using the compass. But we live in a scientific age, and the means exist
to determine and familiarize the people of this country
with the exact variation factor to be applied to our records
of temperature to reduce the expression of heat or cold felt
by human beings everywhere to a common standard of sen-
sibility.

A clothed, living body, having a great evaporating surface through the pores of the skin, is affected by what is
known as the evaporation or sensible temperature; which
is found by placing the thermometer bulb in nearly the
same environment as the human body in summer—that is,
by clothing or surrounding it with cotton, dipping into a
humid source, so that the capillary tubes of the cotton
fibers may carry around the bulb moisture, as perspiration
is carried to the surface of the body through the skin. The
resulting evaporation about the moistened surfaces of the
human body and the thermometer is similar, and the
greater the dryness of the air, the greater and the more
rapid is the evaporation and the resulting coolness. A
gentle wind carries off the layers of air in contact with the
body as they become more or less saturated with moisture,
and they are replaced by drier air, thus promoting evaporation whereby the temperature of the surface is lowered.
Every one has felt the sensation caused by wind blowing on
damp garments or on wet skin, and the cold thus experienced. The normal skin gives off a quantity of water in

the form of perspiration, and in proportion to the dryness of the air this moisture disappears by evaporation. The passage of this moisture into vapor causes the abstraction of heat from the body, and the bodily temperature is lowered, as may be readily observed some little time after severe exertion. Light cotton or linen fabrics allow the perspiration to pass through freely, so that the evaporation and cooling process is unchecked.

The dryness of the arid region is most favorable to these cooling influences, while in the East the close, humid air, being already almost constantly saturated with moisture, is unable to absorb the moisture on the skin; and so not only is there an absence of the cooling effects of evaporation, but the perspiration remaining on the body helps to clog the pores and thus produces the well known and thoroughly uncomfortable suffocating effect.

When the air is saturated with moisture—a condition often present in the East during the heated term—there is absolutely no evaporation; consequently, in such cases, the deduction of our temperature from this cause is zero, and the sensible temperature thermometer and the ordinary thermometer read alike. But this is seldom or never the case in the arid region, on account of its dryness.

The variation between the sensible temperature and the reading of the ordinary thermometer is greatest in the hottest season of the year, and during the hottest part of the day, and that is precisely the time when it is most needed.

As there is a signal service record of the readings of these two kinds of thermometers for a number of years, taken at 7 a.m., 3 p.m., and 11 p.m., I will take, as representing the extreme heat occurrence, the "means" of those readings for the month of July for a period of years for all places of observation in the United States, and compare them by drawing isotherms showing the reading of the sensible temperature thermometer and the ordinary thermometer, and contrast them.

Yuma, Arizona, which is but a few miles from the Gulf of California, and is influenced by the moist winds therefrom, is generally reputed to be the hottest place in the United States. Fortunately, to controvert this, we have a signal service weather record for that point, as we have also of the cities on the Mexican gulf, and on our South Atlantic shore line. From these records it appears that the mean sensible temperature, deduced from the three daily observations for the month of July at Yuma is but 75°. Turn to the East to find where like conditions prevail, and incredible as it may seem, we discover that we have not a single one of the shore line cities between Wilmington, N. C., and Brownsville, Tex., at which the mean July sensible temperature does not exceed this 75° at Yuma. Not only is this true, but all the citrus districts of Florida, the sugar-cane region of Louisiana, and the tobacco lands of Texas, are south of the 75° line, and so are sensibly warmer than Yuma, Arizona.

Yuma, as before stated, is affected by the moist winds blowing from the Gulf of California; therefore its sensible temperature is not as low as many of the valleys (which are susceptible of reclamation by irrigation) in the midst of the so-called deserts of California and Arizona.

As this is one of the startling facts brought out by the investigation of the data upon which this paper is based, permit me to repeat it. The coast of South Carolina and

Georgia, all of Florida, the seaboard of Alabama and Mississippi, nearly the whole of Louisiana, and the southeast part (one-third) of Texas, is not so well favored in July as Yuma, Arizona, which is the most humid place, hence the most uncomfortable perhaps in the arid season.

North of the line of the Yuma or 75° July mean sensible temperature, of which the sections last noted are to the south, lies the belt of sensible temperature between 75° and 70°. The upper edge of this zone or the line of 70° for July, may be located by commencing at Chesapeake Bay, near Washington City, following the eastern foothills of the Alleghany range, turning north at Chattanooga, including West Tennessee and Kentucky, extreme Southern Indiana and Illinois, Southeast Missouri, including the city of St. Louis, following closely the north and northwest boundaries of the Indian Territory and Texas, also Southwest Arizona, and Southeast California.

Having discussed the mean sensible temperature of the warmest month, a glance at what is shown for the warmest part of the day in the hottest month may serve to further accentuate the comparative comfortableness of the arid region. Yuma, Arizona, has a mean sensible July temperature at 3 p.m. of 78°; Charleston, S. C., Titusville, Flor., Galveston, and Brownsville, Tex., have the same; Key West is 1° degree hotter. Phoenix, Arizona, farther from the influence of the moist atmosphere of the Gulf of California, is 4° cooler than Yuma in the hottest part of the day.

It may be said that the average of 11°, the observation including those at 7 in the morning and 11 at night, for the month of July, represents only general conditions and not special instances; but here, also, to controvert this assumption, I have authoritative signal service data. As it is desired to show only the side of the case least favorable to arid America, leaving the genial dry air and sunshine of winter in the arid regions uncontrasted with the cold waves, slush, and humid somberness of the Eastern winters, only midsummer extremes will be stated.

As Yuma is a regular signal service station, where complete records have been kept for twenty years, let us see what are the extremes there. The greatest shade temperature recorded is 118°, but, as this was registered by a self-recording thermometer, the evaporation temperature at the same time is not given. However, at another time, when 116° was recorded, the wet-bulb thermometer was at 70°. It is well known that this dry heat produces no injurious effects, sunstrokes being unknown.

It follows from these recorded facts that in the hottest parts of the arid region the midsummer weather is not only endurable, but even enjoyable and refreshing. Those are the facts as they exist now, when the present conditions—the bare soil, etc.,—are especially conducive to high temperature. But it may be readily conceived that there will take place salubrious modifications, as some of us have already realized, when these desert places are covered with the green carpet of alfalfa and the verdure of trees; when the wasting waters are stored and diverted by the irrigator to the surface of a soil only waiting for water to produce bountifully, not only the fruits of the earth in due season, but almost to produce the seasons themselves at will."

That the difference between the reading of the wet and dry-bulb thermometers in the dry, hot atmosphere of the

Salt River Valley is much greater than is actually experienced by the human body I must admit, but it is equally true that the higher the reading of the dry-bulb goes, the greater is the perspiration, and the more nearly do the conditions of the body conform to that of the wet-bulb, and more nearly are the actual heat conditions experienced by a person registered by this wet-bulb.

The actual heat experienced in this climate by the human body varies from 5 to 20 degrees lower than the reading of the dry-bulb thermometer, and is influenced by the percentage of humidity, by the degree of heat, and the amount and kind of clothing worn.

If the human body could be kept in the same condition of moisture as is the wet-bulb, and in the same strong current of air, the reading of the wet-bulb would accurately register our sensation of heat.

If a person should remove all clothing, wrap himself in a wet sheet and stand out in the sunshine with a stiff wind blowing, those conditions would approximate the conditions of the wet-bulb. As this is not the conventional or convenient mode of dress, it is not practicable, and these conditions are never fully realized. They are more nearly attained by the laboring man in the fields, who is in a constant copious perspiration.

Judging from pretty careful observation I apprehend that under average conditions, if you will divide the difference in the reading between the wet and dry-bulb by two, and add this to the reading of the wet-bulb, you will arrive at the correct decree of heat experienced by the human body in the Salt River Valley.

Our summers are hot. The sunshine is continuous during the day. The nights are cool, comfortable, balmy, almost seductive. If a perfect night is ever experienced it is here during the summer. The heat is stimulating, healthful, and not the least depressing. Perspiration is copious and evaporation instant. One feels well and soon gets to long for the summer time, when people live out of doors both day and night. That lassitude which one feels during the dog days in moister climes is entirely absent.

This is the season when the invalid makes his greatest improvement, when he sleeps with only the sky for a covering, and contentedly swings in his hammock during the day, filling up at his pleasure on luscious fruit in great variety.

I can give you data for the month of May just passed, which I think will surprise you. As you all know this month gave us everywhere a taste of what hot weather is. The following table will give you a slight idea of what summers are like in the Salt River Valley when nature surpasses herself.

TABLE OF ACTUAL AND SENSIBLE TEMPERATURES FOR THE MONTH OF MAY, 1896, WITH THE PERCENTAGE OF RELATIVE HUMIDITY FOR EACH DAY EXCEPT SUNDAY.

(Observations taken at 3:00 P.M. by U. S. W. B., Voluntary.)

DATE.	Actual Dry-Bulb.	Sensible Wet-Bulb.	Per Cent Relative Humidity.	DATE.	Actual Dry-Bulb.	Sensible Wet-Bulb.	Per Cent. Relative Humidity.
May 1--	82.	55.	11	May 17--	---	- -	--
" 2--	85.5	57.7	13	" 18--	89.2	59.	11
" 3 -	---	- --	--	" 19--	84.2	55.	9
" 4--	85.5	57.5	13	" 20--	87.	57.	11
" 5--	85.2	59.	17	" 21--	86.8	58.	14
" 6-	75.4	51.4	13	" 22--	89.	59.5	13
" 7	72.5	51.	18	" 23--	85.9	61.6	22
" 8	73.5	52.	19	" 24--	---	- -	--
" 9--	81.	56.	31	" 25--	104.	67.	12
" 10--	---	---	--	" 26--	100 5	68.	10
" 11--	83.	54.	9	" 27--	108.8	69.	11
" 12-	83.5	56.	9	" 28--	108.5	69.	10
" 13--	85 6	56.	11	" 29--	98.	66.	16
" 14-	87 2	54.5	7	" 30--	87 3	59.4	15
" 15--	85.	55 5	10	" 31--	---	---	--
" 16--	83	54.	9				

This is a rapidly growing community of industrious, in-
telligent, law-abiding people, where the stranger is welcome
and is soon made to feel at home.

The conditions of climate and soil are such that anything that will grow in any other part of this country can
be grown in the Salt River Valley just as readily and in
the majority of instances to much better advantage than
elsewhere. Lands are cheap and easily attained and any-
one who so desires can find profitable employment.

That you may the more readily concede to Phoenix and
vicinity its proper place, at the head of the list, as a city
which leads all others in natural sanitary conditions and
healthfulness, I will here reproduce some vital statistics
taken from the May number of *The Sanitarian*, comprising
the annual death rate per one thousand inhabitants for 1895
in the following cities:

Compare the death rate of Phoenix, Arizona, 5.04, with that of—

Salt Lake City, Utah	7.37	Minneapolis, Minn.	8.96
St. Paul, Minn.	9.86	Buffalo, N. Y.	11.12
Denver, Colo.	10.37	Kansas City, Mo.	13.28
Concord, N. H.	14.05	Milwaukee, Wis.	14.37
Los Angeles, Calif.	15.84	Tampa, Fla.	20.59
Portland, Maine	24.75	Mobile, Ala.	29.44
St. Louis, Mo.	17.07		

These few will serve to make my point; the other prin-
cipal cities of the United States, with their more or less
perfect sanitary conditions, range between Kansas City,
Mo., with her 13.28, and Mobile, Alabama, at 29.44.

As to diseases and their curability or alleviation by the
climatic conditions and surroundings of the Salt River Valley, I have tried to be sufficiently explicit, and will leave you
to draw your own conclusions and make your own deductions.

U. S. DEPARTMENT OF AGRICULTURE—WEATHER BUREAU.
WEATHER DATA AT PHOENIX, ARIZONA, 1896.

(Observations taken at 8 A.M. and 8 P.M. 75th M. time. Corresponds to 5:32 A.M. and 5:32 P.M. local time.)

		January.	February.	March.	April.	May.	June.	July.	August.	September.	October.	November.	December.
Mean temperature	Dry-bulb A.M.	43	44	51	51	61	73	77	77	72	60	48	44
	bulb P.M.	63	67	73	76	87	102	97	98	92	78	67	63
	Wet-bulb A.M.	39	39	44	43	49	58	69	69	63	54	43	39
	bulb P.M.	50	49	52	52	58	67	72	73	69	61	54	49
Extremes	Highest	79	82	92	89	110	115	109	108	104	98	83	75
	Lowest	30	28	34	38	45	61	69	69	55	47	32	33
*Relative humidity	A.M.	69	65	56	50	41	40	68	65	60	70	67	65
	P.M.	40	25	21	15	14	13	30	33	33	41	42	35
Percentage of sunshine		77	87	75	91	89	98	73	85	82	81	81	79
Total rainfall		.46	.05	.39	.05	T	T	4.25	1.77	1.18	1.02	.64	.67
The normal temperature as determined from 13 years' observation		49	54	61	67	75	83	90	88	81	69	58	53
The average rainfall as determined from 16 years' observations		.57	.89	.68	.30	.16	.07	.85	.97	.54	.62	.44	1.12

*Percentage.

W. T. BLYTHE,
Observer and Sec. Director U. S. Weather Bureau.

TABLE XIX.—COMPARATIVE METEOROLOGICAL DATA FOR THE WINTER MONTHS OF 1896-97.

	November, 1896			December, 1896			January, 1897			February, 1897			March, 1897		
	Phoenix	San Diego	Los Angeles	Phoenix	San Diego	Los Angeles	Phoenix	San Diego	Los Angeles	Phoenix	San Diego	Los Angeles	Phoenix	San Diego	Los Angeles
Number of days clear	18	23	14	19	20	15	14	17	13	14	14	11	20	18	11
" " part cloudy	9	3	12	11	8	12	9	7	13	10	2	11	8	4	13
" " cloudy	3	4	4	1	3	4	8	7	5	4	12	6	3	9	7
Percentage of sunshine during month	85			79			63			70			83		
Inches of rainfall during month	$\frac{6.4}{100}$	$1\frac{05}{100}$	$1\frac{04}{100}$	$\frac{67}{100}$	$2\frac{04}{100}$	$2\frac{06}{100}$	$3\frac{10}{100}$	$3\frac{10}{100}$	$3\frac{10}{100}$	$\frac{70}{100}$	$2\frac{70}{100}$	$5\frac{62}{100}$	$\frac{53}{100}$	$1\frac{10}{100}$	$2\frac{10}{100}$
Excess of precipitation above normal	$\frac{20}{100}$	$\frac{20}{100}$	$\frac{27}{100}$		$\frac{25}{100}$		$3\frac{10}{100}$	$1\frac{100}{100}$	$\frac{77}{100}$		$\frac{46}{100}$	$2\frac{36}{100}$			
Deficiency of precipitation below normal				$\frac{1}{100}$		$1\frac{80}{100}$				$-\frac{43}{100}$	$\frac{100}{100}$		$\frac{100}{100}$	$\frac{100}{100}$	$\frac{100}{100}$
Number of rainy days in which $\frac{1}{100}$ of an inch or over, fell	1	5	5	2	5	5	10	9	9	3	9	9	5	8	7
Mean relative humidity, per cent	54	76	72	50	72	66	62	71	68	54	72	74	47	71	74

THE INFLUENCE OF IRRIGATION ON CLIMATE AND HEALTH.

An article by Wm. Lawrence Woodruff, M.D., Phoenix, Arizona, published in *The Irrigation Age*, for August, 1896.

The conclusive discussion of this subject implies a study of the physical conditions of the given locality—a comparison of meteorological data for a considerable period while arid conditions prevailed, with similar data after the same territory has been brought under irrigation—consideration of the percentage of humidity most conducive to health, with the prevailing temperatures, altitude, and wind movements, and the determination of actual and ascertained general effects, as shown by freedom from disease in the community and by vital statistics. Each of these elements of the problem must be studied in its relation to all the others. The inquiry is inherently difficult and complex under the most favorable conditions.

Captain William A. Glassford, Signal Corps, U. S. A., of Denver, Colo., a high authority in such matters, says in a recent article: "In the hottest parts of this arid region the midsummer weather is not only endurable, but even enjoyable and refreshing. These are the facts as they exist now, when the present conditions — the bare soil, etc.— are especially conducive to high temperature. But it may be readily conceived that there will take place salubrious mod-

31

ifications, as some of us have already realized, when these desert places are covered with a green carpet of alfalfa and the verdure of trees; when the wasting waters are stored and diverted by the irrigator to the surface of a soil only waiting for water to produce bountifully, not only the fruits of the earth in due season, but almost to produce the seasons themselves at will."

In the nature of the case we could not expect any definite scientific data for this vicinity prior to the practice of irrigation. The precipitation is about seven inches per annum. Without it settlement and residence are impracticable in a locality in which agriculture must depend for moisture solely upon irrigation. In the Salt River Valley, settlement and irrigation came hand in hand.

The Salt River Valley, with Phoenix as its center, is situated in the vicinity of the 33d parallel of north latitude.

The surrounding physical and climatic conditions are totally different from those of any other locality under irrigation, and must be understood in order to arrive at right conclusions.

It has an elevation ranging from 1,000 to 1,500 feet above sea level.

High mountain ranges surround it on all sides, save on the southwest, where it verges into the larger Gila Valley.

The Gila Valley, under similar conditions, extends to the Gulf of California, which in turn extends with its 53,000 square miles of surface well into the tropical zone.

This great inland sea, with its mouth 250 miles wide, flanked on either side with continuous mountain chains, acts as a funnel into which the tropical waters and winds, sweeping from the equator up the Mexican coast, enter.

These surroundings and winds are largely the influences which go to produce our peculiar and phenomenal climatic conditions.

It is universally conceded that an atmosphere carrying too much moisture is unfavorable to perfect health. It may not be so well known, but is equally certain that the air may be too dry. A couple of my patients had this experience. During a long drive upon the desert, on an exceedingly hot day, the air became extremely dry and fairly burned. Their throats became parched and perspiration ceased. No amount of water taken internally seemed to relieve this condition, which was speedily followed by a languor and then stupor, bordering on coma. This thoroughly alarmed the wiser of them, and sensibly, during the remainder of the day they took turns, fifteen minutes in duration, one driving while the other gratified the irresistible desire to sleep, and in this way they reached irrigated ground in safety. The same phenomena have been observed in numerous other cases. I am satisfied this explains many cases of death upon the desert which have heretofore been attributed to lack of water. During the summer time, in this locality, elimination by the kidneys is reduced one-half. Perspiration is immensely increased and the skin becomes the chief eliminating organ of the system. When the percentage of humidity in the air gets below a certain point, the evaporation from the surface of the body becomes too instant, the surface burns, perspiration and elimination of effete material cease, thus producing the phenomena above described. I attribute these effects entirely to a lack of sufficient moisture in the atmosphere.

I am not prepared, as yet, at least, to fix definitely the

point at which the percentage of moisture in the air is neither too great nor too little. Investigation may, and probably will, show that the most favorable degree of saturation would vary according to individual characteristics. It is probable there is a range of 10 or 12 degrees within which it is difficult, if not impossible, to say that any given point would be more favorable to general health than another. It may be safely said that in the temperature of the Salt River Valley, during the summer, a humidity below 8 per cent. is disadvantageous, while that above 20 per cent. begins to become oppressive.

Both actual and sensible temperature, as shown respectively by the readings of dry and wet-bulb thermometers, must always be considered in connection with the humidity. In every climate there are seasons when the percentage of humidity is excessive, and results generally in a feeling of depression. In the Salt River Valley these periods are usually limited to say a week in February, and a week in August, very much less in duration than in any other locality within my observation. There is very little wind here. The mean average hourly movement at Phoenix for a period of years is stated by the United States Signal Service at two and 37-100 miles. It would be interesting to compare the humidity of the higher lands of the valley near the foothills with that in the lower valley, but no data exist for such statement. We know that it is 10 to 15 degrees warmer in winter and cooler in summer, for instance, on the lands under the Rio Verde Canal on the north side and the Highland Canal on the south side of the Salt River than near the river at Phoenix. The extremes of temperature between day and night are much less on the higher lands than in the lower valley, and the danger of taking cold is proportionately reduced.

· It is probable that the effects of irrigation on climate and health under the high temperature and low relative humidity of this valley are somewhat different from those in regions of lower temperature, greater humidity, and either higher or lower altitude.

It is almost impossible, without accurate observations, to make comparisons, or to arrive at safe, definite conclusions as to the influence of irrigation on climate, either in a general way or in a given locality. I have been unable to procure any data whatever as related to this valley, or to any similar locality, showing the relative humidity before and after irrigation. Without such facts I can only state conclusions arrived at from personal observation and study of its effects on this locality.

I am decidedly of the opinion that upon the deserts of Arizona, without irrigation, the moisture in the atmosphere is sometimes so little as to interfere with health and comfort, and produce feverish conditions. The evaporation of water from the irrigated land supplies this deficiency to the air and obviates the injurious tendencies.

I have frequently had this experience. The "wetting down" of my well-shaded porch on a hot summer day lowers the temperature, as shown by the thermometer hanging upon the wall, 10 to 15 degrees. This result from the refrigeration of the air in the process of evaporation of the water.

It is well known that a well-shaded dwelling in the midst of an alfalfa field is much cooler than the same residence surrounded by bare ground. This is due in part, perhaps, to the absence of reflection from the earth, but chiefly, I think, to a similar slight refrigeration of the air by the evaporation of the moisture in the earth and vegetation of the surrounding field. The effect becomes still more marked when a gentle breeze is blowing.

At Phoenix, during the summer months, the air is so dry that the mid-day registration of relative humidity ranges from 6 to 15 per cent. It rarely goes above the latter point, and if it were not for irrigation it would drop still lower, which is not desirable.

My conclusion is that the evaporation of moisture from irrigated surfaces slightly increases the moisture in the air and promotes the healthfulness of both animal and plant life.

That the evaporation from irrigation has but slight influence in increasing the dampness in the surrounding air will be readily understood when we recall the following facts: That moist air is lighter in weight than is dry air. That moisture is evaporated as an invisible gas. That being lighter and a gas, it is not a disturbing atmospheric element. That it instantly rises with great velocity to a point in the atmosphere where the temperature is below its own dew point, where it becomes visible in the form of clouds. But a very small portion of the evaporated moisture is retained in the lower and warmer strata of air. The hotter the air the greater is the evaporation from the irrigated ground. This evaporation lowers the earth's temperature and also that of the surrounding air.

During the winter months, the temperature ranges much lower, evaporation is much less, and the air is constantly so dry that the slight influence it exerts is scarcely notice-

able. During the last winter the mean relative humidity
was as follows: 1895, Oct. 53 per cent., Nov. 68 per cent.,
Dec. 58 per cent.; 1896, Jan. 54 per cent., Feb. 45 per
cent., Mar. 38 per cent., Apr. 32 per cent, with a rainfall
during these same months of but 2.70 inches.

It is well known that the best qualities of citrus fruits
can only be grown where there is sometimes danger from
frost. This danger in the citrus localities of the Salt River
Valley only exists for say an hour at a time, and that
about sunrise of a frosty morning. The horticulturist is
able by flooding his irrigation ditches with water at this
time to obviate, or lessen, the danger to his fruit. The
water in the ditches will freeze before the fruit or the trees,
and thus the temperature of the surrounding air is raised.
This phenomenon exists all over the district under irrigation, to a greater or less extent, and the extremes of day
and night temperatures are thus modified.

As to the influence of irrigation on the healthfulness of
the inhabitants of an irrigated district I can be more posi-
tive. It is demonstrated by actual experience to be advantageous. Phoenix and the Salt River Valley is the healthiest place in the United States. Next to it comes Salt Lake
City, Utah, also in an irrigated district.

That part of the Salt River Valley north of the Salt
River, west of the Verde, and east of the Agua Fria, covering a territory of 250 square miles and including the city of
Phoenix, of which the population on a conservative basis
for 1895, is placed at 15,000, had for the year named an
annual death rate of 5.04 per one thousand inhabitants.
Salt Lake City during the corresponding year had a death
rate of 7.37. Our death rate for the five summer months

last year was but one-quarter of one per cent. of population, or 2.54 for one thousand inhabitants in the above named territory. With this showing, no one for an instant can imagine that in the least does irrigation militate against health.

On the other hand, I believe that irrigation is a major factor in increasing the healthfulness of a community. It is probable that on account of our favorable climatic conditions this is more emphatically true of the Salt River Valley than any other locality. I refer this fact chiefly to three causes.

Under an irrigation system, properly operated, there are no water holes, or sloughs in which vegetation grows only to decompose and pollute the air. There are no pools of stagnant water to create miasms. The water supply is under man's control, both as to volume and times of distribution. Vegetation is rank and prolific, but grows only where it is desired, and is limited to valuable products. Useless vegetation is discouraged, but should it by chance exist, it rather dries up than rots.

This low death rate is further explained by the constant living in the open air, which we enjoy to its utmost limit.

Irrigation, by promoting the rapid and phenomenal growth of trees and the verdant grass which carpets our lawns, makes a continuous existence out of doors possible and enjoyable for three-quarters of the year.

We live nature's life as nature intended we should live it, and have our reward of unparalleled healthfulness.

I do not believe there is any other place on earth where children are so universally healthy. This is especially true of the summer season. They are marvelously free from "summer complaint," and kindred ailments. I never saw any place where the children thrive as they do in the Salt River Valley.

To quote Captain Glassford again, he says in the same article: "This greater portion of arid America, elevated high above the humid levels of the East, covered with aspects most sublime of the earth, fed with the most invigorating constituents of the atmosphere, will yet be appreciated; and these elements, under the influence of modern civilization, will produce the hardiest and grandest race of men and women who have yet trod the planet. They will create a western empire and become masters of the continent, if not of the world."

SOME CLIMATIC FEATURES OF THE
ARID REGION.

Extracts from a paper by WILLIS L. MOORE, Chief of United States
Weather Bureau, communicated to Fifth National Irrigation
Congress at Phoenix, December 15, 1896, and published by the
Weather Bureau. An excellent aid to an appreciation of the
bearing of the facts shown by the accompanying tables and
charts.

"Under the direction of the Honorable Secretary of Agriculture it was my pleasure, on September 20, 1895, a few
weeks after coming to the head of the Weather Bureau, to
issue instructions to the observers of the weather service
to begin the telegraphing from observation stations of the
readings of the wet-bulb thermometer, more popularly
known as the "sensible" temperature. This is about the
temperature felt by animal life and may be many degrees
below the air temperature, the difference between the two
temperatures depending upon the relative humidity of the
air—the drier the atmosphere the lower the sensible temperature when compared with the air temperature; the
damper the air the higher the sensible temperature. This
will be better understood when it is stated that in case the
air be saturated, the readings of the dry and the wet-bulb
thermometers will be the same and the sensible temperature
and the air temperature will be equal. In the semi-arid
regions of the West the sensible temperature during the
summer months often is 20° to 30° less than the air tem-

perature, which condition is due to the extreme dryness of the atmosphere. In the more humid regions of the eastern part of the country such extreme differences can not occur.

Within the broad confines of the United States there are many, but not all, shades and varieties of climate. One of the questions most frequently asked the Weather Bureau is, "Where shall I find a climate possessing both dryness and equability of temperature?" To this interrogatory reply must be made that the ideal climate as regards equability of temperature and absence of moisture does not exist in the United States, but that the nearest approach to it will be found in the great Southwest, where all shades of dryness, from a rainfall sufficient for successful agriculture, to the aridity of the desert may be found.

The temperature of the Southwest is not equable in the sense of having an extremely small daily range, but, on the other hand, it possesses the quality of uniformity in a greater degree than will generally be found elsewhere, except on the seacoast. The most equable temperature on the globe will be found on the high table-lands and plateaus of the Tropics. Santa Fe de Bogota, in the United States of Colombia, has an average temperature of about 59° for all months of the year, and the range for the entire year is less than is often experienced in a single day in these latitudes.

But while the ideal temperature may be found on the higher elevations of the Tropics, the rainfall is much greater and more continuous than in this country.

The rainfall of the great Southwest varies with location. Less than 200 miles from the Colorado Desert, where the rainfall is practically *nil*, places may be found whose annual average rainfall is as great or greater than any point in the Middle States of the East. Generally speaking, however, the greater portion is dry, using that term as indicating a rainfall considerably less than 20 inches per annum on the average.

The mountainous portions of Arizona and California have an average annual rainfall ranging between 20 and 50 inches, depending somewhat upon the elevation and geographic position, while the lowland portions and the plateaus, especially east of the Sierras, have a rainfall both small in amount and variable in character. The rainfall records of the arid region, and other portions of the United States, are published in the monthly bulletins of the various climate and crop centers, and in more convenient form in the annual data volumes of the Weather Bureau. It is not possible to report upon them in detail here.

The temperature of a place depends chiefly on three conditions, viz., latitude, elevation, and contiguity to large bodies of water. At sea level in the Tropics extreme conditions of heat and moisture, so combined as to produce very great physical discomfort, abound. But even under the equator it is possible to escape the tropical heat of low levels by ascending from 4,000 to 6,000 feet. In the economy of nature there is a certain limit beyond which the two extremes, dryness and equability of temperature, can not co-exist; thus we may find a region so deficient in moisture as to satisfy the requirements of the case, but the very lack of moisture is a condition that facilitates radiation and thus contributes to great extremes of temperature. Regions may be found, as on the lower Nile, where there is a lack of rainfall coupled with a h.gh and moderately uniform temperature. The mean winter temperature of Cairo, Egypt, is 56°; mean summer temperature, 83°; a range from winter to summer of 27°. The mean winter temperature of Phoenix, Ariz., is 52°; mean summer temperature, 87°; a range of 35°. It is by no means difficult to find a counterpart of the far-famed Egyptian climate in the great Southwest.

The dryness of the air and the clearness of the sky are the conditions upon which daily ranges of temperature depend; the greater these, the greater the range of temperature from day to night. While a high summer temperature is characteristic of the Southwest, it is a fact long known to residents of that section, and somewhat imperfectly realized in other portions of the country, that the sensation of heat as experienced by animal life, is not accurately measured by the ordinary thermometer. The sensation of temperature which we usually refer to the condition of the atmosphere depends not only on the temperature of the air, but also on its dryness, the velocity of the wind, and other circumstances. The human organism, when perspiring freely, evaporates the moisture of its surface and thus lowers its temperature. The meteorological instrument that registers the temperature of evaporation, and thus, in a great measure, the actual heat felt by the human body, is the wet-bulb thermometer. The latter, as indicated by its name, is simply an ordinary mercurial thermometer, whose bulb is wetted with water at the time of observation.

Chart I has been constructed to show the average actual and sensible temperature of Weather Bureau stations in the United States for the summer season.

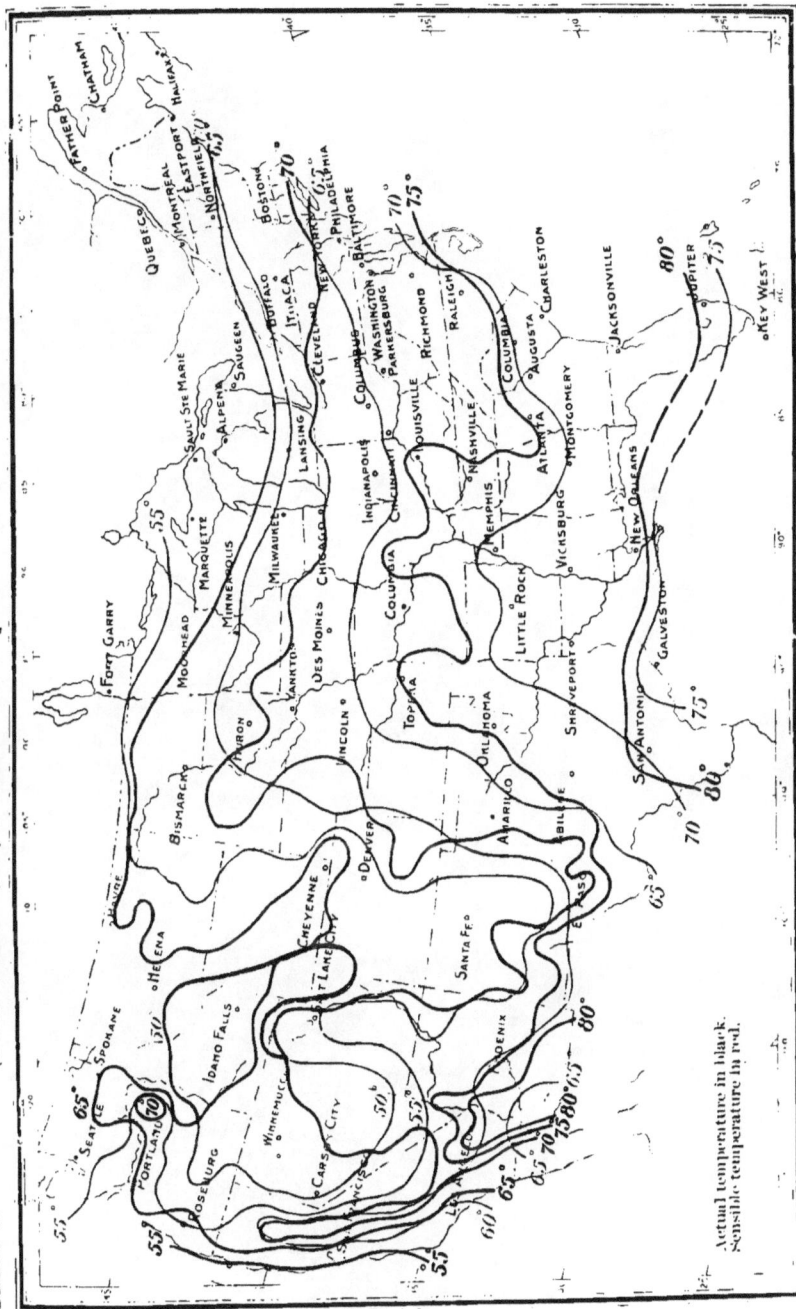

Chart I.

Average Actual and Sensible Temperatures

(Deduced from eight years' observations at 8 a. m. and 8 p. m., 75th meridian time.)

Actual temperature in black.
Sensible temperature in red.

The broad principle illustrated by this chart is that the greatest differences between shade and sensible temperatures are round where the air is the driest, and the least where the air is most humid. A glance at the chart is sufficient to show the general trend of the lines of equal air and sensible temperatures. The great interior valleys and the plains east of the foothills of the Rocky Mountains are uniformly heated under the insolation of summer to an average of from 65° on the northern boundary, to about 80° on the Gulf Coast. The northern portion of this vast extent of country is, moreover, in the path of atmospheric disturbances that pass from west to east over our northern boundaries, thus causing an indraught of warm, moist air from lower latitudes. Again, the distribution of atmospheric pressure over the eastern two-thirds of the United States is at times such as to cause a more or less complete stagnation of the generally eastward drift of the air; the surface of the ground warms up under intense insolation, and loses but little heat by radiation at night; the winds are light southerly or southeasterly and there is an absence of vertical interchange between the warm surface air and the cooler air aloft. Such conditions sometimes extend over the entire Mississippi Valley and eastward to the Atlantic seaboard. On the other hand while it is possible for a heated term to prevail over an arid region by day, the relatively great radiation by night lowers the temperature to an endurable degree, and there is but little bodily discomfort. The heat of the daytime, moreover, is borne without distress by reason of the great dryness of the air. The red lines of Chart I show the temperature of evaporating surfaces in summer in the United States. It will be seen that the line of 60°, which marks the temperature of evaporation of the region of New England and the Great Lakes, passes almost due north and south along the eastern foothills of the Rocky Mountains, and skirts Southern New Mexico and Arizona. The line of 55° passes almost due south from Eastern Montana to Southeastern New Mexico, and thence northwesterly. The temperature of evaporation in all of the territory above this line (55°), embracing almost two-thirds of the arid region ss *below* 55°; in fact, in almost one-third of the region it is not over 50°. The sensible temperature of two-thirds of the United States, or east of the one hundred and fifth meridian, ranges from 55° to 75°. West of the one hundred and fifth meridian the range is from 50° to 65°.

Chart II has been prepared to illustrate the extreme differences that prevail in midsummer, the 8 p.m., seventy-fifth meridian time observation of July having been used. (8 p. m., seventy-fifth meridian, corresponds to 7 p.m. central, 6 p.m. mountain, and 5 p.m. Pacific time.) There is an objection to the use of synchronous time in depicting climatic elements that have a marked diurnal period. Observations taken at the same moment of local mean time should be used whenever possible, but the exigencies of a service instituted for the purpose of forecasting weather changes demand the use of synchronous time. As regards the data of this chart (II), it may be urged with propriety that a comparison of thermometric readings made at the same moment of time from the Atlantic to the Pacific, is misleading, since an accurate estimate can not be made of

Chart II.   Mean Actual and Sensible Temperatures, July (8 p. m.), 75th Meridian Time.

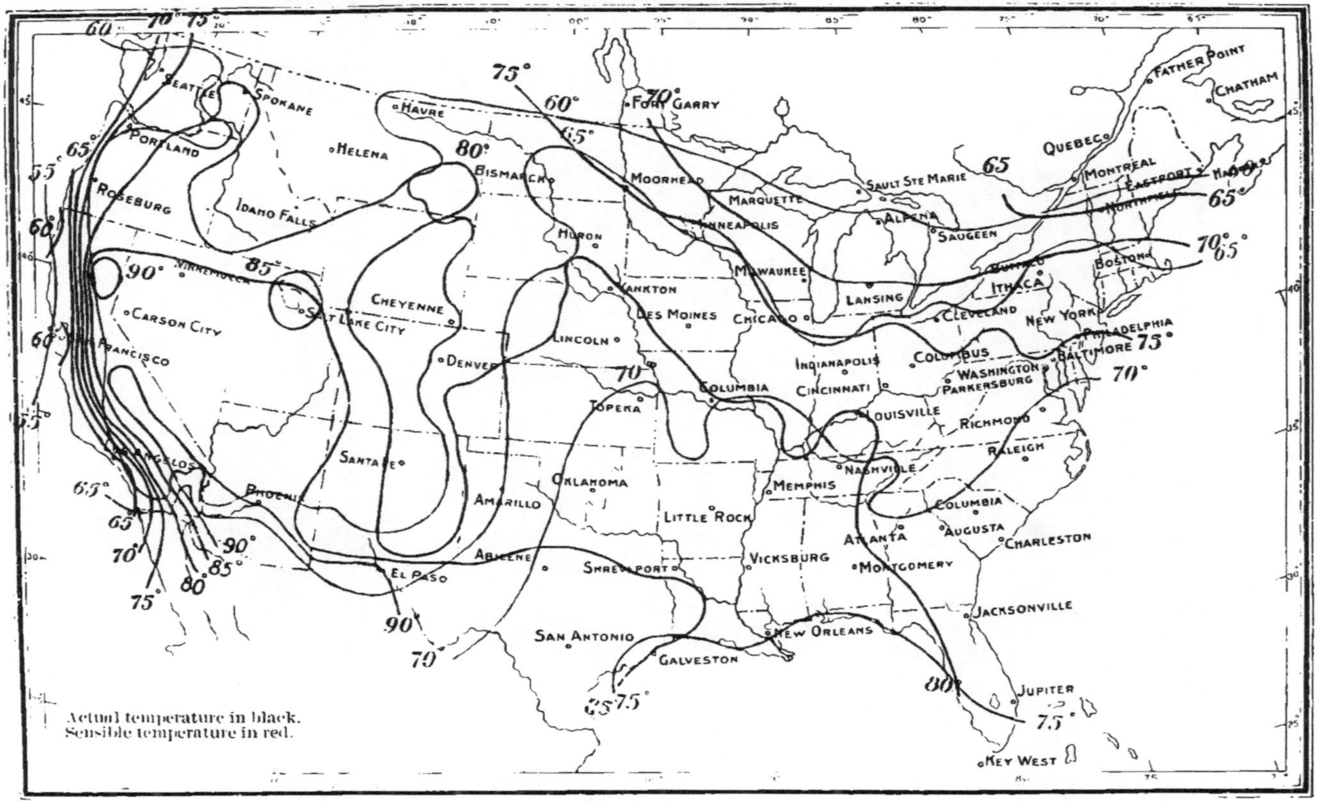

Actual temperature in black.
Sensible temperature in red.

the amount of increase of temperature for western stations due to diurnal influences alone, and it was mainly with a view of illustrating this fact that the chart was prepared.

The thermometer readings on the Atlantic seaboard are made near the hour of 8 p.m., local mean time; those on the California coast are made near 5 p.m., local mean time. Naturally, the Pacific Coast temperatures are considerably higher than those on the other side of the continent, three hours later in the afternoon. The contrast between the two sides of the country is plainly shown by the black lines of equal actual temperature on Chart II, and it will also be observed that the Southwest is the warmest part of the United States.

The lines of equal sensible heart, on the other hand, show an entirely different condition as regards the location of greatest heat. The arid region is now the coolest part of the United States, judged from the temperature of evaporation only. The line of 60° sensible temperature, starting in New England, skirts the northern boundary as far as the one hundred and tenth meridian; thence it follows a south-southeasterly course to Southeastern New Mexico; thence westerly to the neighborhood of Los Angeles, Cal., and thence northerly, with a few unimportant deflections, to the North Pacific coast.

The decrease of temperature from the hour of maximum heat to nightfall is not regular, nor does it bear any definite relation to an increase in longitude reckoned westward from Greenwich. A comparison of the normal 8 p.m. seventy-fifth meridian time temperatures with the normal maximum temperature of the day shows that on the eastern coast line the temperature at 8 p.m. is, on the average, 8° to 12°

lower than at the time of greatest daily heat. In the lake region and lower Ohio Valley the difference is from 5° to 8°. In the upper Mississippi and Missouri valleys and Texas and the plains region the difference averages from 4° to 7°; that is to say, the temperatures at the 8 p.m. observation (corresponding to about 6:30 p.m. local time) are from 4° to 7° lower than the highest point reached by the thermometer during the day. On the eastern slope of the Rocky Mountains, although the evening observation is made at 6 p.m., local time, two hours nearer the time of greatest heat than at New York and Philadelphia, the difference is as great as at the last-named places. In other words, the temperature falls as much by 6 p.m. at Denver, as it does by 8 p.m. in New York and Philadelphia. This would seem to be the result of the greater daily range and more rapid rate of cooling at elevated stations. West of the Rockies the differences range from zero, at Red Bluff, to less than 4° in the great interior basin, and from 5° to 6° in Southern Arizona.

The local vicissitudes of temperature are well illustrated in the case of Red Bluff, Cal., where the average temperature at about 5 p.m., local time, is but four-tenths of a degree below the maximum of the day. Curiously enough, at Los Angeles, in the lower part of the State, the 5 p.m. temperatures are about 10° lower on the average than the maximum of the day.

Chart III has been constructed to show the relative humidity of the United States in summer. The data used in preparing the chart were the synchronous observations at 8 a.m. and 8 p.m., seventy-fifth meridian time, during the eight years 1889-96. The chart itself shows better than

Chart III.

Mean Relative Humidity—Summer.

(From observations at 8 a. m. and 8 p. m., 75th meridian time.)

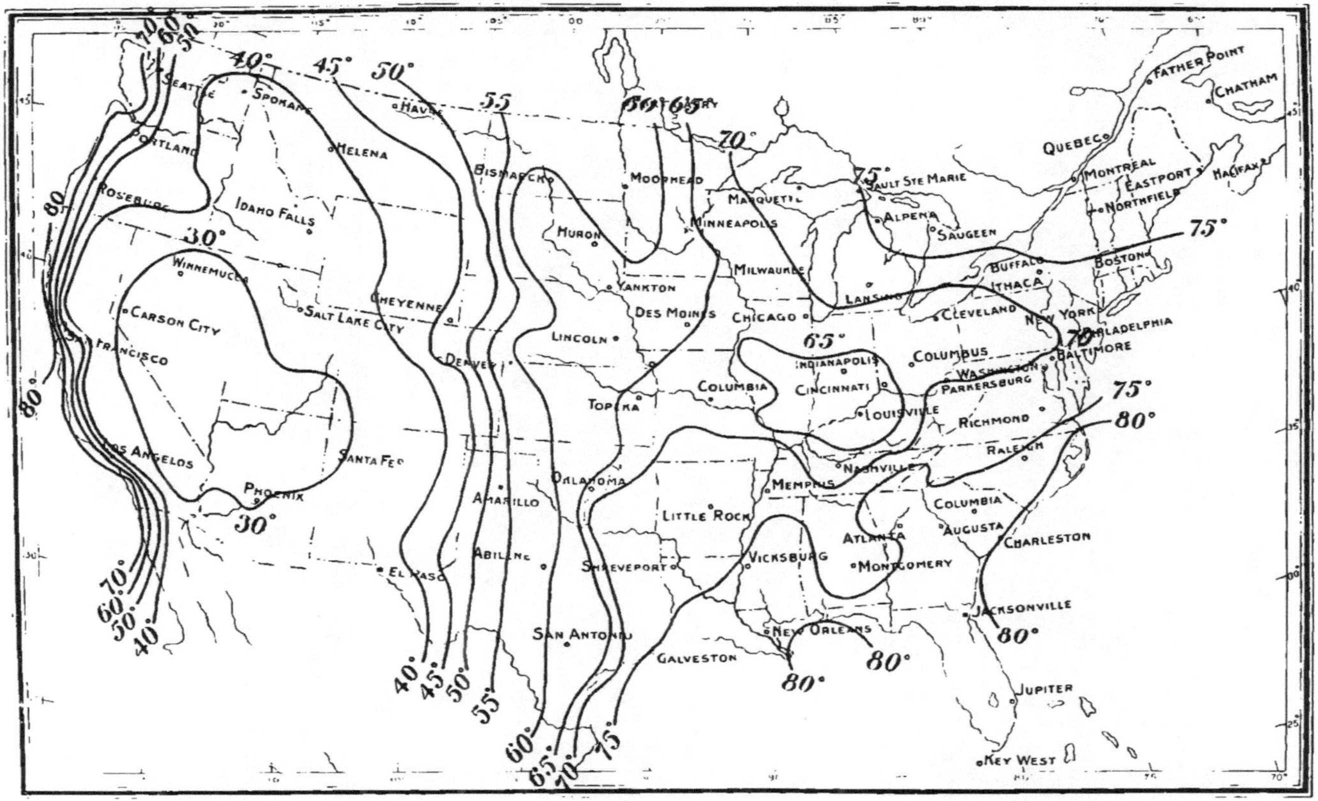

mere words the distinctively dry and humid regions. The influence of the ocean is seen on both coasts, as also that of the Gulf of Mexico and the great lakes.

Broadly speaking, the variation of insolation from day to night, and from season to season, with the changing declination of the sun, is the great controlling agent of climate. The most regular, and at the same time the simplest, climate of the world, is that of the Tropics, where the succession of changes from day to day are as monotonous in their regularity as they are enervating on the human system. The great life zone, the seat of business enterprise and activity, is found in temperate climates. Here the simple diurnal changes of the Tropics are largely masked by irregular changes, the result of the passage of cyclonic and anti-cyclonic systems. The sum total of these changes constitutes the weather of the temperate zone.

Between the Tropics and the temperate zone there are, in certain longitudes, considerable areas where the climate is more or less transitional between the two strongly marked zones. The southwestern part of the United States may be classed as having a climate between the extremes of the Tropics and the temperate zones. Not being within the path of storm frequency, the sequence of weather is more uniform than in more northern latitudes, or on the same parallel farther east. The rainfall is deficient; there is an absence of clouds; insolation by day and radiation by night, are both strong; the range of temperature from day to night is large, from 25° to 35°, depending upon the elevation and character of the surface of the ground; the winds are generally light and the evaporation is high.

TABLE XVII.

Deaths in 1000 inhabitants, 1896.	July.	August.
Phoenix, Ariz.	9	3
Boston, Mass.	$19\frac{54}{100}$	--------
New York City, N. Y.	$26\frac{01}{100}$	$27\frac{09}{100}$
Philadelphia, Pa.	--------	--------
Atlantic City, N. J.	$6\frac{80}{100}$	$11\frac{4}{10}$
Washington, D. C.	$29\frac{03}{100}$	$24\frac{14}{100}$
Charleston, S. C.	--------	--------
Jacksonville, Fla.	--------	--------
Atlanta, Ga.	--------	--------
Tampa, Fla.	--------	--------
Mobile, Ala.	$24\frac{72}{100}$	$22\frac{80}{100}$
Vicksburg, Miss.	--------	--------
New Orleans, La.	$23\frac{40}{100}$	$25\frac{00}{100}$
Little Rock, Ark.	--------	--------
Galveston, Tex.	--------	--------
San Antonio, Tex.	--------	--------
Memphis, Tenn.	$25\frac{61}{100}$	$26\frac{13}{100}$
Cincinnati, Ohio	$17\frac{64}{100}$	$16\frac{32}{100}$
Pittsburgh, Pa.	$18\frac{93}{100}$	$21\frac{30}{100}$
Buffalo, N. Y.	$18\frac{55}{100}$	$16\frac{11}{100}$
Cleveland, Ohio	$22\frac{07}{100}$	$15\frac{26}{100}$
Detroit, Mich.	$19\frac{14}{100}$	$15\frac{88}{100}$
Chicago, Ill.	$18\frac{83}{100}$	$16\frac{5}{10}$
St. Paul, Minn.	$15\frac{04}{100}$	$7\frac{08}{100}$
Des Moines, Iowa	--------	--------
St. Louis, Mo.	20	$21\frac{03}{100}$
Kansas City, Mo.	$13\frac{23}{100}$	$10\frac{54}{100}$
Omaha, Neb.	$5\frac{06}{100}$	$6\frac{00}{100}$
Los Angeles, Cal.	$11\frac{38}{100}$	$11\frac{40}{100}$
San Diego, Cal.	--------	--------

THE SUMMER CLIMATE OF PHOENIX.

An article by WM. LAWRENCE WOODRUFF, M.D., Phoenix, Arizona,
published in *The Medical Century*, for September, 1896.

The month of June, 1896, will be remembered as having
the highest range of temperature, and for the greatest number of consecutive days ever known in the Salt River Valley, if not in the United States.

The following table shows the actual heat as marked by
the reading of the dry-bulb thermometer, the so-called sensible temperature (as indicated by the wet-bulb), and the
relative humidity or percentage of saturation, according to
the observations of the United States Weather Bureau, at
Phoenix, Arizona:

Date.	Actual Temperature in Degrees.	Sensible Temperature in Degrees.	Rel. Humidity. per ct.	Date.	Actual Temperature in Degrees.	Sensible Temperature in Degrees.	Rel. Humidity. per ct.
1	97.1	65.6	16	16	114.0	73.0	12
2	95.9	64.0	12	17	112.5	70.3	10
3	94.0	61.0	11	18	108.0	73.0	17
4	91.0	60.8	14	19	102.0	66.5	12
5	93.8	61.0	12	20	103.5	64.0	7
6	94.8	63.8	15	21	105.0	67.0	10
7	97.0	64.5	13	22	104.2	64.8	8
8	100.8	65.4	12	23	107.0	74.0	20
9	104.8	64.8	8	24	99.2	63.8	11
10	107.0	67.0	9	25	102.0	68.5	16
11	109.0	67.8	7	26	98.4	69.4	21
12	109.5	68.8	10	27	102.2	70.2	19
13	114.8	72.0	13	28	102.8	65.3	13
14	114.5	73.0	11	29	103.2	66.3	11
15	114.0	71.5	10	30	104.0	68.0	13

From June 9th to 18th inclusive was the longest continuous period of extremely hot weather within the memory of
the oldest inhabitant. From the 13th to the 17th, the best
accredited thermometers (set nearer the ground than the
government instrument), registered from 3° to 5° degrees
higher, and indicated from 118° to 120° Fahrenheit. It
will be noted that the difference between the actual and
sensible temperature (indicated by the readings of the dry
and wet-bulb respectively) was from 30° to 43° degrees, depending principally upon the percentage of humidity. On
only seven days did the relative humidity go above 13 per
cent. This is a fair index of the dryness of the summer
air in the Salt River Valley.

With this record of intense heat, extending over one-
third of the month, should be coupled that of the wonder-
ful exemption from disease during the same period. Nowhere else in the known world were the inhabitants so
healthy as in Phoenix and its vicinity. There was practically no acute sickness.

The following table of deaths for June, 1896, in that
portion of the Salt River Valley north of the Salt River,
west of the "Rio Verde," and east of the "Agua Fria,"
containing a population of 16,000 and including the city of
Phoenix, is a fair index of our ordinary summer healthfulness:

Cause of Death.	No. Cases.	Age.	Remarks.
Puerperal fever	1	27	
Typhoid pneumonia	2	28-7	
Bowel disease	1	2	
Typhoid fever and chronic alcoholism	1	79	
Chronic alcoholism and heat prostration	1	64	Tramp.
Old age	2	85-86	
Brain fever	1	24	
Consumption	4		All transients.

In all, thirteen deaths. If the five cases of transients be deducted there are left eight deaths in a population of 16,000 during the hottest month in the history of the community.

During the months of May, June, July, August, and September, 1895, there was but one death each month from bowel trouble among children in the territory named.

During the five summer months of the past four years the total death rate was as follows:

1892, one-fourth of one per cent.
1893, two-fifths of one per cent.
1894, one-third of one per cent.
1895, one-fourth of one per cent.

An average of 2 and 85-100 in 1,000 inhabitants. This is the season, in all other parts of the world, of greatest fatality from gastro-enteric diseases.

Were it possible the world ought to know, not only that the Salt River Valley, during the summer time, is the healthiest spot on earth, but that the healthy individual and the health-seeker can live here in comfort and with pleasure during the heated term. We feel better, brighter, stronger, and have better appetites than in the winter sea-

son. As soon as the weether begins to warm up, aches,
pains, and discomforts vanish. Life is not only livable,
but we live more of life as nature intended we should live it.
  We live in the open air. The lawn is parlor, sleeping
apartment, and often dining-room. The diet is largely
fruit in abundance and of great variety. The foliage of
quick-growing trees forms a grateful shield from the perpetual sunshine of the day, and at night the beauty of the
moonlight is unsurpassed. It is the luxury of life to live in
the open air throughout the dewless night, dressed in the
lightest garments, and without a fear of taking cold. There
could be no nobler canopy than Arizona's clear, blue star-
lit sky. There is rarely a night so warm as to interfere
with sleep.
  The days are hot and the air is dry. One needs to drink
water frequently and copiously. This natural appetite can
be fully gratified without risk. The effect is a profuse perspiration, "flushing" out with it all effete material from the
system. As soon as this perspiration reaches the surface
it is evaporated, and the heat of the body thereby reduced.
This process of refrigeration and elimination is kept up
without interruption for months at a time, and is the expla-
nation of our unparalleled healthfulness.
  This is the period when the invalid makes his greatest
improvement. To get the most benefit from this climate,
he must come during the spring and summer, rather than
in the fall or winter. This is so with the great majority of
cases, the contrary is the exception. It is perfectly safe
for our people from any part of the country to come to the
Salt River Valley during the summer. Our hot, dry air is
stimulating and not in the least debilitating. We usually

find (when there is sufficient vitality left to expect any benefit at all) a gain in weight and strength so long as the hot weather lasts. A summer spent here with its unloading of poisonous, effete, broken-down tissues, prepares an invalid to get the greatest benefit from our genial winters.

TABLE I.—MAXIMUM TEMPERATURE FOR JULY, 1896.

Date	1	2	3	4	5	6	7	8	9	10	11	12	13	14	15	16	17	18	19	20	21	22	23	24	25	26	27	28	29	30	31
Phoenix, Ariz.	104	102	104	104	106	106	104	106	106	104	108	108	108	102	100	96	96	96	96	96	92	94	100	100	96	100	102	102	102	100	102
Boston, Mass.	88	88	84	62	66	62	62	76	84	90	86	92	92	80	86	82	74	74	72	78	86	82	82	72	74	84	80	80	84	84	78
New York, N.Y.	82	80	84	78	82	78	78	76	76	82	82	86	86	86	82	80	74	74	72	78	82	80	76	72	76	82	84	82	88	88	80
Philadelphia, Pa.	84	86	80	80	88	86	80	72	88	82	90	92	92	88	88	82	78	74	82	80	84	80	84	76	80	82	90	88	92	92	86
Atlanta, Ga.	88	88	82	88	90	84	70	74	82	82	84	84	86	88	88	88	70	82	82	84	86	82	90	90	90	94	92	94	94	96	94
Washington, D.C.	86	86	90	88	90	84	84	74	84	82	86	90	90	88	88	88	80	80	82	82	86	82	82	76	80	84	94	92	92	92	92
Charleston, S.C.	84	84	88	86	86	86	84	84	84	86	86	88	90	90	88	88	84	82	86	88	88	88	88	76	92	96	96	90	94	92	94
Jacksonville, Fla.	90	86	88	88	88	86	86	80	84	88	88	90	94	92	88	92	88	88	88	90	90	94	92	94	94	96	96	98	96	100	100
Atlantic City, N.J.	76	74	76	74	84	72	72	72	76	76	84	88	82	82	78	76	72	74	74	76	70	72	84	72	78	78	80	90	88	90	82
Tampa, Fla.	88	90	86	90	88	86	84	86	76	76	86	88	90	82	78	80	88	74	74	76	70	72	84	88	90	90	90	90	92	88	90
Mobile, Ala.	90	86	84	90	90	84	80	86	88	84	86	88	88	92	90	90	90	88	86	86	86	88	90	92	90	90	90	90	94	94	98
Vicksburg, Miss.	94	96	94	92	92	90	90	86	88	92	84	92	92	94	92	92	94	94	94	92	92	94	92	92	94	94	94	94	94	94	98
New Orleans, La.	90	84	84	94	92	90	90	86	90	88	90	88	90	90	90	90	88	84	84	88	86	92	88	92	88	92	92	92	92	92	96
Little Rock, Ark.	94	94	96	94	88	88	90	82	82	88	88	86	94	96	96	94	98	98	96	94	92	96	100	102	100	100	100	100	100	100	102
Galveston, Tex.	88	86	86	86	86	90	94	90	88	88	82	86	86	86	88	88	88	88	88	88	88	96	88	88	88	88	88	88	88	88	90
San Antonio, Tex.																															
Memphis, Tenn.	90	94	94	92	88	88	86	86	84	86	86	88	92	92	94	92	86	94	92	88	88	90	92	96	88	96	96	96	96	100	102
Cincinnati, Ohio	90	84	88	84	82	84	80	76	76	84	88	88	88	88	90	76	74	84	82	78	80	84	82	84	80	90	94	92	96	96	84
Pittsburg, Pa.	88	90	86	84	78	76	78	70	76	84	86	70	88	84	80	74	80	76	84	80	86	80	76	78	78	80	82	84	82	88	80
Buffalo, N.Y.	78	84	82	82	78	76	68	78	72	74	76	76	76	82	78	74	74	76	74	78	78	78	74	64	70	74	84	84	86	86	74
Cleveland, Ohio	84	86	82	82	78	78	74	74	72	80	84	84	86	82	78	70	72	80	80	78	78	78	72	66	74	80	82	82	86	86	76
Detroit, Mich.	84	86	86	84	72	70	76	78	70	80	86	88	92	82	80	72	76	76	74	78	82	82	78	68	76	74	84	84	90	88	76
Chicago, Ill.	86	86	84	76	68	70	68	80	72	84	90	90	90	94	80	64	72	74	74	84	78	84	78	68	78	92	88	88	92	88	72
St. Paul, Minn.	88	88	84	78	82	80	78	76	72	86	86	94	92	82	74	72	70	84	82	84	78	84	78	72	88	92	82	82	82	84	78
Des Moines, Iowa	80	88	86	78	82	86	78	78	82	86	90	90	90	90	82	76	80	74	82	84	86	78	70	74	76	94	86	86	92	90	78
St. Louis, Mo.	88	88	88	88	82	86	78	80	84	86	88	88	90	94	96	82	84	86	88	82	80	90	92	82	82	90	98	86	96	98	94
Kansas City, Mo.	80	92	96	78	82	84	80	76	84	84	88	88	90	94	90	78	84	84	80	84	88	90	92	66	84	94	82	90	90	92	90
Omaha, Neb.	82	96	84	76	84	80	82	76	80	84	88	92	90	94	98	80	78	74	80	84	90	80	72	72	78	94	82	86	92	88	84
Los Angeles, Cal.	78	82	80	76	70	78	80	82	82	86	80	80	92	88	88	88	82	82	76	82	80	84	80	76	78	76	76	78	78	84	88
San Diego, Cal.																															

TABLE II.—MINIMUM TEMPERATURE FOR JULY, 1896.

Date	1	2	3	4	5	6	7	8	9	10	11	12	13	14	15	16	17	18	19	20	21	22	23	24	25	26	27	28	29	30	31	Nights as hot or hotter than at Phoenix.
Phoenix, Ariz.	74	72	75	70	80	80	76	76	70	80	78	78	78	78	72	70	70	70	70	74	72	76	76	76	70	70	76	76	78	78	76	
Boston, Mass.	60	66	68	68	56	58	58	58	66	74	72	70	72	72	66	66	60	60	64	64	70	72	72	60	56	60	68	68	70	68	64	2
New York, N.Y.	66	64	64	64	62	68	68	64	62	66	70	70	72	72	66	66	60	60	64	66	70	70	70	66	62	66	68	68	70	70	68	6
Philadelphia, Pa.	62	66	68	70	76	72	72	66	64	72	72	72	72	78	74	74	58	62	64	66	70	72	72	66	66	66	68	72	74	70	68	7
Atlanta, Ga.	68	68	68	68	72	70	68	60	60	68	68	72	68	70	70	74	68	68	66	68	68	72	68	76	74	74	74	74	76	74	78	5
Washington, D.C.	58	60	70	68	74	72	72	64	64	68	72	68	74	76	74	74	62	56	60	70	70	74	72	66	70	60	70	74	72	72	72	20
Charleston, S.C.	76	70	74	78	72	74	74	72	74	72	72	74	72	72	76	76	74	74	78	78	78	78	78	80	80	80	80	80	80	82	82	15
Jacksonville, Fla.	74	72	74	74	74	74	72	72	74	72	74	74	74	76	72	72	—	—	74	76	74	74	76	74	76	74	74	74	78	78	72	2
Atlantic City, N.J.	64	64	68	70	68	68	68	68	64	68	68	72	70	70	72	66	62	60	64	64	64	68	66	66	66	64	66	66	70	66	74	14
Tampa, Fla.	77	72	72	70	70	74	70	72	70	70	74	74	70	72	74	74	74	72	70	72	72	70	74	74	74	74	70	72	72	78	80	18
Mobile, Ala.	76	74	74	74	74	74	68	70	72	72	72	72	72	76	74	74	70	70	74	70	74	76	74	76	76	76	74	74	76	74	74	18
Vicksburg, Miss.	70	74	74	70	70	72	72	70	68	70	72	76	74	74	74	74	76	74	76	74	76	76	76	76	76	76	76	78	78	78	78	22
New Orleans, La.	78	70	76	74	74	76	74	74	76	74	76	76	76	74	76	76	76	78	76	76	76	76	74	76	70	76	76	78	78	78	78	18
Little Rock, Ark.	70	70	72	72	72	72	72	66	66	72	74	72	74	74	74	76	78	78	78	78	78	76	76	76	76	76	76	76	78	80	80	26
Galveston, Tex.	80	80	76	76	72	74	78	80	78	80	78	80	80	78	82	82	82	82	82	80	80	82	82	82	82	80	80	76	80	80	80	13
San Antonio, Tex.	72	74	74	74	72	72	74	74	74	76	72	70	72	72	72	76	76	76	76	76	74	74	74	72	74	72	74	74	74	74	74	17
Memphis, Tenn.	70	70	72	74	72	72	74	74	74	76	74	74	74	74	70	70	72	72	72	72	72	72	72	72	74	72	74	74	78	78	74	6
Cincinnati, Ohio	62	68	68	70	70	68	62	60	60	66	66	66	66	66	70	66	56	58	68	68	74	72	72	70	70	64	74	74	76	80	80	5
Pittsburg, Pa.	62	66	70	72	70	68	62	58	58	62	62	64	70	68	72	62	52	58	64	62	70	72	72	58	62	62	66	72	66	66	64	1
Buffalo, N.Y.	62	68	70	72	70	62	62	56	58	60	60	62	70	64	64	54	56	56	64	66	66	70	56	60	64	64	68	70	72	60	60	3
Cleveland, Ohio	62	66	68	70	66	62	56	60	60	66	66	70	70	68	70	66	56	56	56	68	68	70	56	60	56	68	68	70	70	76	68	1
Detroit, Mich.	64	70	68	70	62	62	62	60	58	58	66	68	68	72	68	54	52	60	68	66	66	70	60	56	56	62	66	70	72	72	62	4
Chicago, Ill.	68	72	70	70	70	62	60	54	54	66	66	68	68	72	68	78	56	62	64	64	70	68	66	58	60	66	66	70	70	70	70	0
St. Paul, Minn.	66	66	68	58	66	66	54	56	50	62	64	68	72	68	72	56	56	60	62	64	66	68	66	54	54	54	60	58	58	64	62	1
Des Moines, Iowa	68	68	68	58	58	62	52	50	50	54	58	64	66	60	56	60	56	66	66	66	64	64	64	58	58	56	66	64	68	72	72	12
St. Louis, Mo.	70	72	72	70	70	66	62	60	62	60	64	74	74	74	68	66	58	62	74	74	74	68	70	76	66	72	78	80	80	80	76	7
Kansas City, Mo.	70	72	74	70	62	64	62	60	64	64	66	66	70	70	72	66	66	68	68	70	70	74	72	60	58	76	76	74	72	70	76	3
Omaha, Neb.	68	68	68	64	64	64	64	58	60	62	60	68	68	68	68	70	62	62	64	68	68	64	58	56	56	70	66	66	70	70	64	0
Los Angeles, Cal.	56	54	58	58	60	58	60	60	60	62	62	62	64	64	64	62	64	60	60	64	64	—	64	60	62	60	58	—	54	56	58	0
San Diego, Cal.	58	58	60	58	60	62	62	62	62	64	64	64	64	66	66	68	66	64	64	66	68	68	64	66	66	66	64	—	58	56	60	0

TABLE III.—TEMPERATURE 8 A.M., (DRY-BULB THERMOMETER) FOR JULY, 1896.

Date	1	2	3	4	5	6	7	8	9	10	11	12	13	14	15	16	17	18	19	20	21	22	23	24	25	26	27	28	29	30	31	Local Time	Days as hot or hotter than at Phoenix.
Phoenix, Ariz.	74	72	78	82	80	86	80	76	84	82	78	80	78	82	72	70	76	70	76	76	72	74	76	82	70	76	76	76	78	78	76	5:32	
Boston, Mass.	70	74	75	58	66	60	66	72	74	77	76	78	82	76	72	66	64	68	68	70	76	74	74	68	60	68	70	76	76	68	70	8:16	6
New York, N.Y.	64	63	68	61	74	72	72	66	60	72	77	79	76	74	76	68	62	68	66	70	74	70	74	66	64	68	70	72	76	74	74	8:04	2
Philadelphia, Pa.	72	72	72	74	78	74	74	68	78	78	76	74	74	82	78	74	66	64	70	74	74	70	78	66	64	68	72	72	76	74	74	8:	13
Atlanta, Ga.	74	72	70	68	74	74	70	68	78	76	76	74	74	82	72	74	66	64	72	74	74	74	78	68	68	72	78	78	80	78	76	7:20	13
Washington, D.C.	71	70	74	72	76	74	74	64	74	74	70	76	78	78	72	76	66	68	70	68	76	76	76	72	70	76	76	76	80	80	80	7:52	13
Charleston, S.C.	80	80	80	82	82	76	80	80	72	80	80	80	80	78	80	80	78	82	80	78	82	80	80	84	82	82	82	84	84	84	84	8:	26
Jacksonville, Fla.	78	76	78	78	82	78	78	64	78	82	80	80	78	80	78	76	78	80	78	78	80	78	78	78	78	78	78	78	80	78	78	7:30	22
Atlantic City, N.J.	70	68	80	80	78	78	76	68	72	74	74	76	80	76	74	76	66	76	74	70	68	78	78	70	74	74	68	80	78	78	78	8:04	11
Tampa, Fla.	78	78	78	78	78	78	78	76	72	80	80	78	78	78	80	82	80	78	78	78	80	78	78	82	80	80	78	80	82	84	78	7:25	24
Mobile, Ala.	82	80	74	72	74	78	72	60	74	74	74	76	78	76	76	78	78	74	76	74	80	80	78	82	80	80	78	78	82	80	80	7:10	18
Vicksburg, Miss.	76	78	80	72	74	74	74	72	74	74	76	76	76	76	76	76	74	80	76	74	80	80	80	78	78	80	78	78	80	80	80	6:50	19
New Orleans, La.	80	80	76	80	80	78	78	78	78	78	78	80	80	76	78	80	80	78	76	78	76	76	76	80	80	78	76	82	80	80	80	7:	22
Little Rock, Ark.	72	74	78	78	76	72	74	68	66	70	70	72	74	76	74	80	72	74	76	80	78	80	80	80	80	78	74	76	76	82	82	6:35	17
Galveston, Tex.	80	80	78	80	82	80	80	82	80	82	80	82	82	82	78	84	82	82	84	84	82	84	84	84	84	82	80	80	82	80	82	6:41	28
San Antonio, Tex.	74	76	76	76	76	74	76	78	78	78	78	78	76	76	74	76	76	78	78	76	76	76	76	74	74	74	74	76	76	76	76	6:20	15
Memphis, Tenn.	72	74	78	76	76	74	74	72	70	70	72	74	74	76	78	80	72	74	76	76	72	74	72	72	66	74	80	80	80	82	72	7:	16
Cincinnati, Ohio	68	74	70	72	72	64	64	64	68	66	66	70	70	76	72	70	60	60	72	74	74	66	60	72	66	63	76	80	80	80	72	7:22	8
Pittsburg, Pa.	66	76	73	74	72	62	62	64	64	64	58	60	64	74	75	64	62	66	66	68	70	74	72	70	66	70	76	70	76	76	68	7:40	5
Buffalo, N.Y.	70	74	74	76	66	68	58	56	68	68	70	72	74	70	74	58	66	66	66	68	70	74	60	62	64	70	72	70	76	76	66	7:45	3
Cleveland, Ohio	70	70	76	66	64	68	68	60	66	66	68	70	72	70	70	66	62	66	64	66	68	72	66	58	64	66	72	70	74	74	70	7:33	3
Detroit, Mich.	70	76	70	64	68	62	62	58	60	64	70	72	74	70	70	58	60	66	70	72	72	62	58	60	60	68	72	74	74	78	64	7:28	3
Chicago, Ill.	74	74	72	62	62	58	58	64	68	70	72	74	72	72	78	58	64	64	66	68	70	68	62	62	66	68	70	74	72	78	70	7:10	4
St. Paul, Minn.	70	68	68	68	62	66	62	60	62	66	64	68	72	72	70	56	56	64	62	64	60	60	58	50	56	66	62	58	72	66	64	6:48	—
Des Moines, Iowa	70	68	68	68	62	66	62	60	64	68	68	72	72	78	74	62	58	66	68	66	68	70	68	60	60	60	68	60	68	70	72	6:50	1
St. Louis, Mo.	76	74	76	74	70	76	64	64	68	70	74	78	76	78	80	66	62	68	70	76	76	74	76	82	68	76	80	82	82	80	72	6:59	13
Kansas City, Mo.	70	76	78	70	64	68	66	66	64	63	68	72	72	74	80	68	66	72	72	70	74	76	74	60	64	78	78	74	72	80	72	6:42	10
Omaha, Neb.	68	76	70	68	64	74	68	62	64	66	70	72	72	76	64	68	64	68	70	74	64	64	68	56	62	72	68	68	72	72	72	5:	3
Los Angeles, Cal.	58	58	60	60	60	60	60	62	64	62	62	62	64	64	64	58	64	64	66	66	66	66	64	62	62	62	58	—	56	58	58	5:09	—
San Diego, Cal.	62	62	60	62	64	64	64	64	64	66	66	62	62	66	68	68	68	66	68	68	68	68	66	66	66	66	66	—	62	58	62	5:	—

TABLE IV.—8 A. M. SENSIBLE (OR WET-BULB) TEMPERATURE FOR JULY, 1896.

Date	1	2	3	4	5	6	7	8	9	10	11	12	13	14	15	16	17	18	19	20	21	22	23	24	25	26	27	28	29	30	31	Days as hot or hotter than at Phœnix.
Phoenix, Ariz.	62	56	61	78	70	74	68	70	70	72	68	74	70	72	68	68	72	68	74	70	72	72	70	68	68	64	66	70	68	68		
Boston, Mass.	60	66	66	56	56	60	58	57	74	70	71	66	68	68	70	60	56	62	64	72	70	56	72	66	60	60	64	68	70	68	60	11
New York, N.Y.	62	62	68	62	72	70	72	60	74	72	70	72	72	70	74	66	58	64	64	68	72	70	66	62	62	62	64	70	72	70	68	16
Philadelphia, Pa.	62	66	70	70	72	72	70	60	74	72	68	72	70	74	74	68	58	62	64	72	74	74	70	60	60	64	68	72	74	72	66	18
Atlanta, Ga.	70	70	70	66	70	72	70	64	60	70	68	72	72	68	74	68	56	62	66	66	68	74	74	74	74	68	74	74	74	74	66	18
Washington, D. C.	66	66	70	70	72	72	70	66	72	70	70	70	72	68	72	70	58	62	66	66	68	72	70	74	74	74	74	70	70	72	68	20
Charleston, S. C.	76	74	76	78	74	76	74	74	74	74	76	74	76	76	78	76	78	76	76	76	74	74	76	78	76	78	78	76	78	80	78	31
Jacksonville, Fla.	76	72	79	74	74	74	74	74	72	70	70	72	76	74	74	74	74	74	74	74	74	74	76	74	74	76	76	76	76	76	74	29
Atlantic City, N. J	62	66	70	68	74	70	76	74	70	74	76	72	72	72	72	76	76	58	68	72	68	72	66	66	62	66	66	76	76	76	68	16
Tampa, Fla.	76	76	74	72	74	74	76	72	74	72	76	76	74	74	74	70	76	76	76	74	70	76	76	76	74	76	78	76	78	78	82	30
Mobile, Ala.	70	74	74	70	72	72	70	68	68	72	72	74	74	74	72	74	72	72	74	74	76	76	74	76	76	74	74	74	74	74	74	29
Vicksburg, Miss.	70	74	74	70	72	70	66	62	68	72	72	74	74	71	74	74	74	74	74	72	74	76	78	76	78	76	74	74	74	74	74	25
New Orleans, La.	76	76	74	76	74	76	76	76	72	72	70	72	74	72	72	72	70	74	74	74	70	70	72	72	72	72	72	70	70	70	72	30
Little Rock, Ark.	66	72	76	76	76	70	70	70	70	74	74	78	72	74	72	74	74	74	74	74	74	76	72	76	70	72	70	70	70	70	72	24
Galveston, Tex.	76	76	78	76	76	76	76	76	74	74	78	76	74	74	72	72	72	74	74	74	74	74	74	76	70	78	76	76	76	76	78	31
San Antonio, Tex.	70	72	74	72	72	74	72	72	72	72	70	72	70	74	72	72	66	70	74	74	74	74	74	74	72	72	74	74	74	74	72	29
Memphis, Tenn.	68	70	74	76	72	72	76	72	70	62	62	70	72	72	72	70	66	70	72	72	72	72	74	72	72	78	76	72	72	72	72	24
Cincinnati, Ohio	60	65	72	70	62	58	58	56	58	62	66	70	70	72	74	66	54	60	68	70	70	72	66	70	62	62	72	76	76	76	68	14
Pittsburg, Pa.	63	66	69	70	68	62	58	63	64	66	66	68	70	70	70	56	56	60	64	70	70	72	66	60	60	64	72	70	72	68	66	14
Buffalo, N. Y.	62	66	63	70	62	62	54	58	64	66	66	66	66	66	68	50	56	56	62	66	68	72	56	58	56	56	62	68	68	72	58	7
Cleveland, Ohio	60	58	68	70	62	60	56	56	60	60	60	62	62	66	70	56	56	56	56	62	70	70	70	58	62	64	70	72	72	70	64	7
Detroit, Mich.	62	58	68	70	62	60	56	56	60	58	64	64	64	66	68	54	54	54	64	64	64	70	56	60	58	62	66	70	70	72	56	7
Chicago, Ill.	62	64	66	70	62	58	62	54	58	60	64	64	64	66	68	54	64	64	66	60	68	70	76	58	66	66	70	70	70	74	66	9
St. Paul, Minn.	60	61	68	58	60	54	54	56	58	62	62	62	62	62	70	52	56	60	62	58	62	68	62	58	58	66	66	58	56	56	56	5
Des Moines, Iowa	61	64	68	68	60	62	56	56	54	58	66	60	66	70	58	50	52	60	62	60	58	56	56	56	54	54	58	62	66	62	56	2
St. Louis, Mo.	68	70	72	74	68	66	60	60	60	64	66	70	66	72	56	56	52	64	68	66	66	60	70	72	58	70	66	72	66	70	68	18
Kansas City, Mo.	61	70	72	70	70	60	62	54	62	62	62	62	68	70	74	64	66	58	70	72	74	72	72	58	60	60	62	68	68	74	74	14
Omaha, Neb.	66	74	68	66	66	58	58	60	58	60	60	64	66	68	72	60	70	70	68	88	70	58	64	56	60	70	60	62	68	64	70	9
Los Angeles, Cal.	56	56	58	58	60	58	60	60	60	62	62	62	64	64	64	64	64	64	64	64	64	64	64	64	60	60	60	60	--	54	54	1
San Diego, Cal.	58	60	58	60	60	60	62	64	62	62	62	64	64	64	64	64	64	64	64	64	66	64	64	64	64	64	60	--	58	56	60	1

TABLE V.—8 A. M., RELATIVE HUMIDITY FOR JULY, 1896.

Date	1	2	3	4	5	6	7	8	9	10	11	12	13	14	15	16	17	18	19	20	21	22	23	24	25	26	27	28	29	30	31	Days more humid than av. in cit
*Phoenix, Ariz.	50	35	35	84	61	57	54	74	49	62	60	54	85	55	100	90	82	90	82	91	91	91	82	55	90	66	52	59	67	60	66	17
Boston, Mass.	56	67	66	97	94	89	93	63	61	74	76	55	56	66	80	69	57	70	64	71	91	86	81	50	92	59	71	68	76	60	54	21
New York, N.Y.	90	71	100	90	91	91	100	71	100	100	91	91	82	82	91	90	79	81	90	90	91	100	82	77	90	77	72	91	82	98	74	17
Philadelphia, Pa.	57	73	91	82	91	91	82	71	83	75	91	54	82	62	83	74	53	98	72	91	83	100	67	63	63	65	82	75	75	75	59	27
Atlanta, Ga.	82	91	100	90	82	91	100	100	91	91	90	91	91	82	91	91	83	100	100	100	91	100	91	91	91	83	83	83	83	83	75	21
Washington, D.C.	78	75	81	90	85	87	93	81	93	92	85	77	70	75	79	78	61	74	80	80	81	93	72	77	70	77	82	69	85	66	70	21
Charleston, S.C.	83	83	81	84	69	76	91	83	100	75	83	75	83	83	100	83	83	76	83	83	76	83	68	83	76	84	82	77	84	77	77	21
Jacksonville, Fla.	91	82	75	75	83	76	83	83	82	75	83	71	83	83	91	83	83	83	83	83	83	83	75	83	83	84	83	83	76	83	83	24
Atlantic City, N.J.	64	90	97	92	100	100	100	80	82	74	91	61	68	82	91	100	61	64	65	82	90	100	75	81	64	66	91	83	83	83	75	21
Tampa, Fla.	91	91	83	75	52	83	83	82	100	82	83	83	83	83	75	76	83	83	82	83	83	91	76	75	75	83	83	76	84	76	92	22
Mobile, Ala.	76	75	91	75	83	83	100	83	82	91	91	61	83	91	83	76	83	91	83	75	83	83	83	84	91	83	83	83	83	83	75	21
Vicksburg, Miss.	74	83	75	91	91	91	82	73	57	74	91	91	83	91	83	82	75	83	91	83	83	91	100	100	91	75	83	83	75	75	75	23
New Orleans, La.	83	83	91	100	75	83	83	75	71	75	91	68	91	75	75	61	83	83	91	83	83	83	83	68	75	75	75	76	75	61	75	23
Little Rock, Ark.	91	82	75	91	91	91	74	75	71	72	91	76	91	91	75	91	91	83	75	75	82	68	75	68	83	75	67	67	61	61	62	19
Galveston, Tex.	83	83	100	100	83	76	83	76	75	69	92	76	76	76	83	76	69	76	75	83	76	76	69	69	69	84	76	69	83	83	84	21
San Antonio, Tex.	82	82	91	91	82	82	82	82	82	83	100	91	100	91	91	82	81	83	83	91	91	91	91	75	91	91	91	91	82	91	91	25
Memphis, Tenn.	82	82	83	91	82	82	82	75	75	72	82	91	91	91	75	82	81	82	83	82	85	75	83	75	91	75	75	68	62	62	91	22
Cincinnati, Ohio	63	66	100	100	83	57	70	60	75	80	82	91	82	83	83	90	68	71	82	82	100	83	71	75	80	91	75	68	83	83	100	19
Pittsburg, Pa.	82	62	82	87	91	93	97	74	76	65	72	82	84	91	92	82	76	75	72	82	81	87	71	97	78	81	82	93	85	93	90	20
Buffalo, N.Y.	64	66	74	74	80	81	78	88	76	71	81	73	74	91	82	56	69	53	80	90	90	82	77	79	60	64	82	90	74	74	61	17
Cleveland, Ohio	64	73	91	86	73	81	88	63	90	64	57	66	91	91	81	60	68	60	82	82	91	78	78	100	100	89	90	82	81	72	74	18
Detroit, Mich.	55	66	90	100	90	63	79	63	89	100	57	65	74	90	90	78	68	71	72	100	82	91	69	68	79	82	90	83	91	83	60	19
Chicago, Ill.	50	58	73	91	79	80	59	61	77	61	55	58	74	75	75	67	60	51	90	90	64	90	71	89	61	90	81	82	91	83	81	17
St. Paul, Minn.	81	81	90	79	79	80	78	89	77	90	82	91	82	91	70	65	77	79	100	79	89	78	89	88	88	89	79	89	65	80	60	22
Des Moines, Iowa	72	81	100	100	89	86	69	78	78	79	77	73	82	74	74	79	90	90	90	79	63	90	89	89	89	100	79	71	81	80	82	22
St. Louis, Mo.	66	82	82	91	74	81	79	79	63	72	74	75	83	83	75	90	79	90	91	91	91	91	82	76	90	91	75	69	69	76	69	23
Kansas City, Mo.	72	74	91	91	79	81	81	61	70	72	71	82	91	82	75	81	71	79	91	90	89	78	71	89	79	67	60	82	82	68	75	22
Omaha, Neb.	72	91	90	90	79	58	69	61	70	71	72	73	82	74	79	79	70	100	90	90	82	70	100	100	79	91	63	77	82	65	82	23
Los Angeles, Cal.	89	89	89	89	100	89	100	79	90	100	100	100	90	90	100	100	100	90	90	90	90	90	100	89	89	89	89	88	87	87	89	24
San Diego, Cal.	79	89	89	89	89	89	100	90	86	90	80	80	90	90	81	90	81	90	81	81	90	90	90	90	90	90	71	79	89	89	89	21

* During July, 1896, the rainfall at Phoenix was 4.25 inches, an excess over the average for seventeen years, of 3.13 inches, the most humid July in the cit[y] history. Relative humidity is indicated by the percentage of moisture in the atmosphere, complete saturation being indicated by 100.

TABLE VI.—8 P. M., TEMPERATURE (DRY-BULB THERMOMETER) FOR JULY, 1896.

Date	1	2	3	4	5	6	7	8	9	10	11	12	13	14	15	16	17	18	19	20	21	22	23	24	25	26	27	28	29	30	31	Local Time.
Phoenix, Ariz.	102	102	103	100	104	97	88	105	104	94	94	107	101	99	91	96	90	89	95	91	85	94	99	78	94	98	99	100	101	106	101	5:32
Boston, Mass.	78	76	76		60	58	61		78	79	79	78	78	72	80	70	72	66	67	72	80	75	72	66	71	71	74	77	73	80	68	8:16
New York, N. Y.	75	73	76	72	71	72	71	65	72	74	80	83	79	72	80	70	72	67	67	75	77	73	72	66	72	73	74	77	78	80	72	8:16
Philadelphia, Pa.	76	76	82	81	78	78	73	68	78	81	81	84	85	84	83	73	73	71	72	82	79	83	76	73	77	77	73	84	86	86	71	8:04
Atlanta, Ga.	85	79	74	82	79	71	67	71	78	78	81	78	80	74	77	75	76	68	72	80	83	74	86	86	87	84	84	80	90	92	88	8:
Washington, D. C.	79	80	73	79	79	73	65	65	78	74	74	78	83	77	82	69	69	75	75	82	83	74	75	73	74	80	80	80	78	81	73	7:20
Charleston, S. C.	78	78	80	80	75	76	76	76	77	80	80	81	84	74	84	84	77	78	81	80	81	82	79	84	85	80	82	84	87	79	76	7:52
Jacksonville, Fla.	75	80	77	73	77	77	78	74	74	80	82	85	82	84	82	82	75	76	76	84	80	79	83	83	80	82	81	75	85	84	79	7:40
Atlantic City, N. J	75	82	76	84	81	78	82	82	81	76	82	83	83	84	76	83	72	77	78	82	83	83	85	84	85	85	84	85	86	83	87	7:30
Tampa, Fla.	81	80	77	73	77	77	76	74	74	74	75	77	77	76	77	79	79	77	75	73	79	79	79	78	78	79	81	81	79	80	79	8:04
Mobile, Ala.	81	82	76	84	81	78	78	82	81	76	82	82	83	84	76	83	73	77	78	82	83	83	85	84	84	78	84	85	86	87	87	7:25
Vicksburg, Miss.	78	78	79	73	73	80	74	63	82	74	75	82	85	88	85	86	86	78	77	84	84	85	85	85	83	85	78	82	86	80	79	7:10
New Orleans, La.	79	81	77	80	88	81	85	82	81	83	82	82	91	83	83	83	87	87	83	84	84	87	85	90	83	85	79	82	91	90	90	6:50
Little Rock, Ark.	86	91	88	80	81	85	79	79	79	83	78	82	91	93	83	86	88	91	88	91	88	88	90	96	92	93	96	94	95	96	99	7:
Galveston, Tex.																																6:35
San Antonio, Tex.	95	94	94	82	77	83	86	91	92	78	74	76	80	84	88	87	88	84	89	86	88	89	92	90	90	91	92	92	89	92	92	6:41
Memphis, Tenn.	85	90	90	85	75	83	80	76	79	83	78	88	88	91	88	79	82	86	87	86	86	88	86	92	86	92	93	93	93	95	95	6:20
Cincinnati, Ohio	86	77	82	73	73	81	84	72	70	80	84	86	86	82	71	71	73	81	79	74	79	82	76	72	87	87	89	89	91	80	81	7:
Pittsburg, Pa.	73	71	77	76	68	79	69	63	68	79	75	82	82	78	76	71	77	77	78	75	79	73	73	74	74	78	79	81	84	71	75	7:22
Buffalo, N. Y.	73	81	88	71	68	72	67	74	74	72	72	72	72	78	69	66	71	77	78	73	74	73	67	62	72	73	74	74	73	76	71	7:40
Cleveland, Ohio	79	72	73	78	71	74	70	70	63	77	82	70	70	77	60	70	63	77	69	78	75	74	65	65	71	76	75	78	72	76	72	7:45
Detroit, Mich.	80	74	72	72	70	72	72	72	62	82	82	73	73	79	66	64	66	72	69	76	78	79	65	65	72	73	78	80	84	75	70	7:33
Chicago, Ill.	79	84	74	66	68	62	68	70	69	83	85	89	76	90	61	63	66	65	69	83	77	80	79	65	72	73	78	73	83	70	72	7:28
St. Paul, Minn.	76		85	74	80	77	76	76	80	81	84		80	92	77	74	75	72	79	82	68	70	60		75	88	82	81	87	85	68	7:10
Des Moines, Iowa	83	79	86	76	79	72	73		79	82	84	87	87	92	77	74	75	73	79	82	68	70	60	77	85	88	81	81	87	85	68	6:48
St. Louis, Mo.	83	79	86	76	79	72	73	76	79	82	85	87	87	90	77	74	75	72	79	82	68	70	60	77	90	90	82	81	91	85	78	6:50
Kansas City, Mo.	90	90	76	75	80	81	76	74	79	81	85	86	87	90	74	71	71	73	75	83	86	83	88	67	83	89	85	88	91	89	91	6:59
Omaha, Neb.	80	93	83	72	83	67	76	71	78	81	85	87	87	93	74	78	70	70	78	82	80	73	62	71	91	91	81	82	91	84	63	6:42
Los Angeles, Cal.	70	71	70	66	65	69	67	74	74	75	76	83	82	96	76	75	76	74	71	73	71	76	76	73	69	69	70	71	73	75	75	5:
San Diego, Cal.																																5:09

TABLE VII.—8 P. M. SENSIBLE (OR WET-BULB) TEMPERATURE FOR JULY, 1896.

Date	1	2	3	4	5	6	7	8	9	10	11	12	13	14	15	16	17	18	19	20	21	22	23	24	25	26	27	28	29	30	31	Days as hot or hotter than at Phoenix.
Phoenix, Ariz.	65	65	68	73	71	70	72	72	72	74	71	71	73	74	75	75	72	71	71	75	75	71	69	71	72	72	69	72	72	68	71	
Boston, Mass.	67	71	56	58	57	69	58	60	65	71	67	72	70	68	70	64	61	60	61	74	71	57	63	61	63	64	70	70	70	68	53	6
New York, N. Y.	68	70	75	75	69	69	71	65	57	71	67	72	70	71	70	64	60	60	64	73	74	73	72	66	63	66	70	72	73	78	62	11
Philadelphia, Pa.	67	69	73	74	73	74	68	62	74	71	70	70	73	72	74	61	60	64	64	74	74	75	64	71	62	65	71	74	73	78	60	13
Atlanta, Ga.	71	72	69	73	72	70	65	64	69	66	73	74	74	71	72	72	73	67	73	74	74	72	77	76	78	77	75	74	76	77	60	19
Washington, D. C.	69	69	70	74	74	71	71	66	64	71	71	73	74	74	80	65	61	63	68	75	74	74	64	71	66	69	80	76	75	70	61	17
Charleston, S. C.	—	—	—	—	—	—	—	—	—	—	—	—	—	—	—	—	—	—	—	—	—	—	—	—	—	—	—	—	—	—	—	—
Jacksonville, Fla.	74	75	74	75	72	73	72	72	71	75	74	72	74	74	77	74	74	73	75	75	75	75	74	76	76	73	71	76	75	74	72	—
Atlantic City, N. J	74	75	74	75	72	73	73	72	71	75	74	72	74	72	74	74	73	73	75	75	75	75	74	76	76	73	75	75	78	74	72	24
Tampa, Fla.	73	75	74	71	74	74	73	73	73	75	75	74	72	74	73	75	74	74	74	73	74	74	76	78	76	76	75	78	80	76	—	24
Mobile, Ala.	76	75	75	74	75	75	70	72	74	73	73	76	73	76	74	77	72	76	76	75	75	76	77	78	74	75	77	76	76	78	77	20
Vicksburg, Miss.	69	73	74	74	73	74	71	66	61	68	72	73	73	73	73	76	74	73	73	73	75	75	77	75	75	75	74	74	74	74	73	24
New Orleans, La.	74	76	75	75	73	73	75	69	74	76	76	76	75	75	76	76	70	75	75	75	76	76	77	76	75	77	75	75	77	74	80	29
Little Rock, Ark.	70	73	74	71	74	75	72	64	65	69	74	76	74	75	77	70	74	76	75	74	74	76	77	74	74	76	73	75	75	74	74	23
Galveston, Tex.	72	72	72	76	71	73	72	74	77	73	72	72	74	75	70	73	74	75	74	74	75	75	72	73	73	72	71	76	73	73	73	—
San Antonio, Tex.	72	75	75	78	71	73	74	77	68	66	73	72	76	75	76	73	76	76	76	76	75	78	75	79	77	77	75	74	74	75	77	23
Memphis, Tenn.	71	75	75	78	71	74	69	68	66	70	73	73	76	76	76	73	73	76	76	75	74	74	75	76	79	77	75	74	74	75	77	26
Cincinnati, Ohio	69	66	70	75	70	68	60	61	59	66	77	69	74	76	70	60	59	66	70	74	74	74	72	67	67	77	76	79	79	72	70	14
Pittsburg, Pa.	69	70	71	73	70	68	69	59	59	67	69	73	73	68	68	58	59	65	70	73	71	71	61	69	64	68	71	75	70	66	66	9
Buffalo, N. Y.	64	64	69	69	64	64	63	59	64	64	65	68	69	59	68	58	58	62	68	67	70	71	61	64	64	68	66	67	66	66	56	1
Cleveland, Ohio	65	70	73	72	70	65	61	62	59	66	66	68	69	72	62	58	56	58	68	72	72	72	61	64	64	70	73	74	74	70	62	8
Detroit, Mich.	69	71	74	70	69	68	69	60	60	74	69	70	70	73	59	55	56	63	66	70	72	69	59	60	66	69	71	75	74	66	62	6
Chicago, Ill.	66	72	71	64	62	57	56	58	66	62	65	71	68	74	55	50	60	61	68	71	68	61	58	61	62	77	73	71	77	64	64	7
St. Paul, Minn.	—	—	—	—	—	—	—	—	—	—	—	—	—	—	—	—	—	—	—	—	—	—	—	—	—	—	—	—	—	—	—	7
Des Moines, Iowa	68	78	75	64	66	62	60	59	61	65	68	72	72	74	65	59	63	70	71	70	66	61	58	61	68	77	68	68	77	73	66	—
St. Louis, Mo.	73	77	78	74	63	63	67	64	63	68	72	72	72	79	74	62	67	70	74	77	75	78	72	79	73	76	80	77	78	80	75	8
Kansas City, Mo.	73	77	70	69	68	68	70	64	62	65	65	70	72	75	70	65	70	74	74	74	78	78	76	73	76	76	80	77	78	79	67	18
Omaha, Neb.	73	75	73	63	67	62	62	61	62	66	71	72	72	75	64	66	66	70	72	73	73	62	60	63	69	78	69	73	78	75	67	16
Los Angeles, Cal.	63	64	63	62	61	64	63	61	62	66	67	68	70	70	66	63	68	66	64	66	70	66	65	63	69	61	62	62	63	63	64	10
San Diego, Cal.	—	—	—	—	—	—	—	—	—	—	—	—	—	—	—	—	—	—	—	—	—	—	—	—	—	—	—	—	—	—	—	0

TABLE VIII.—8 P.M., RELATIVE HUMIDITY FOR JULY, 1896.

Date	1	2	3	4	5	6	7	8	9	10	11	12	13	14	15	16	17	18	19	20	21	22	23	24	25	26	27	28	29	30	31	Days more humid than at Phoenix.
Phoenix, Ariz.	11	11	15	9	18	25	46	19	20	38	31	16	26	31	47	38	44	41	31	48	64	32	21	70	34	27	20	25	25	16	21	
Boston, Mass.	56	78	78	90	87	100	98	72	81	69	52	60	68	79	71	79	77	67	65	90	76	82	40	100	66	60	84	69	86	88	34	31
New York, N. Y.	70	84	86	90	89	95	72	88	96	91	66	62	63	65	76	74	57	82	88	93	80	95	63	61	61	66	98	62	81	72	59	31
Philadelphia, Pa.	61	72	67	72	77	83	69	83	83	94	58	49	56	90	80	44	44	67	62	77	68	91	67	91	63	66	63	63	66	65	43	30
Atlanta, Ga.	50	71	80	65	68	75	93	62	62	55	49	54	73	90	80	49	44	67	71	67	91	77	91	93	66	68	76	50	63	52	51	30
Washington, D. C.	60	56	88	81	79	93	95	97	93	88	77	83	73	90	80	85	87	95	70	72	89	85	83	93	64	69	50	83	87	58	50	31
Charleston, S. C.	—	—	—	—	—	—	—	—	—	—	—	—	—	—	—	—	—	—	—	—	—	—	—	—	—	—	—	—	—	—	—	—
Jacksonville, Fla.	83	85	75	78	84	87	91	82	74	79	75	66	63	90	74	63	87	79	80	78	76	71	77	69	65	73	76	66	67	79	79	30
Atlantic City, N. J	—	—	—	—	—	—	—	—	—	—	—	—	—	—	—	—	—	—	—	—	—	—	—	—	—	—	—	—	—	—	—	—
Tampa, Fla.	91	73	85	91	85	84	85	93	95	81	74	60	62	63	65	72	93	87	91	63	72	77	73	94	94	76	74	91	73	84	81	31
Mobile, Ala.	79	94	94	60	78	67	63	72	85	83	83	76	81	69	92	77	98	89	91	78	71	71	68	71	60	64	70	63	70	82	64	31
Vicksburg, Miss.	70	80	79	91	92	90	80	75	58	76	88	87	81	85	81	80	79	79	91	95	85	83	100	85	85	83	81	82	77	77	75	31
New Orleans, La.	70	80	89	91	50	70	72	42	66	79	80	76	60	55	66	71	67	89	66	66	69	80	70	70	73	66	71	71	67	53	65	30
Little Rock, Ark.	44	42	52	66	72	63	62	45	47	49	83	80	45	41	75	63	52	50	59	50	45	58	56	36	43	45	33	39	39	36	30	30
Galveston, Tex.	—	—	—	—	—	—	—	—	—	—	—	—	—	—	—	—	—	—	—	—	—	—	—	—	—	—	—	—	—	—	—	—
San Antonio, Tex.	32	34	34	78	75	63	51	45	36	80	88	80	77	57	41	50	54	67	52	56	52	53	37	44	50	40	36	49	46	40	40	28
Memphis, Tenn.	52	50	48	73	66	66	59	50	51	51	59	56	50	57	49	66	54	67	60	64	63	53	60	50	73	51	47	41	40	39	44	29
Cincinnati, Ohio	40	79	70	78	86	84	43	52	52	81	50	59	58	50	49	52	66	67	60	64	69	64	82	79	60	61	47	41	40	39	58	29
Pittsburg, Pa.	47	64	73	85	74	58	57	57	60	53	52	66	81	76	100	62	44	54	71	89	80	91	53	83	56	65	86	76	80	95	64	31
Buffalo, N. Y.	60	38	57	91	81	65	83	40	59	68	52	66	72	87	100	62	50	54	42	81	87	88	64	82	68	74	69	76	84	59	36	31
Cleveland, Ohio	47	91	72	73	68	62	60	40	59	69	69	54	95	78	55	46	47	42	56	81	78	88	64	74	68	74	69	86	84	59	36	31
Detroit, Mich.	56	86	72	93	95	82	78	48	97	91	57	52	86	76	39	55	53	61	86	93	73	73	60	82	68	84	73	81	81	62	57	31
Chicago, Ill.	48	54	83	90	71	76	46	48	44	31	33	41	86	76	66	60	59	82	95	59	74	36	70	75	45	62	65	88	64	66	66	31
St. Paul, Minn.	67	73	62	60	48	43	38	35	32	40	45	49	47	43	52	40	51	89	67	54	93	60	92	56	70	61	62	51	63	72	65	29
Des Moines, Iowa	98	77	70	89	55	95	52	44	39	48	46	69	77	62	69	50	59	73	82	91	87	90	75	78	78	61	39	51	56	56	90	29
St. Louis, Mo.	73	56	75	74	58	58	50	44	41	48	51	49	49	47	83	50	59	73	82	91	87	95	75	71	76	55	63	64	59	63	87	31
Kansas City, Mo.	72	44	63	60	42	71	40	56	41	50	50	46	48	60	61	57	95	100	93	65	70	80	65	64	70	55	64	64	66	63	47	31
Omaha, Neb.	72	44	63	60	42	71	40	56	30	44	50	46	48	44	61	58	38	93	68	80	70	55	89	64	70	55	51	64	56	66	93	29
Los Angeles, Cal.	68	70	68	80	80	76	78	66	66	66	66	46	55	45	61	51	66	66	68	68	81	74	61	66	72	65	66	62	59	51	56	30
San Diego, Cal.	—	—	—	—	—	—	—	—	—	—	—	—	—	—	—	—	—	—	—	—	—	—	—	—	—	—	—	—	—	—	—	—

TABLE IX.—MAXIMUM TEMPERATURE FOR AUGUST, 1896.

Date	1	2	3	4	5	6	7	8	9	10	11	12	13	14	15	16	17	18	19	20	21	22	23	24	25	26	27	28	29	30	31
Phoenix, Ariz.	102	106	94	98	98	100	98	102	106	106	102	106	108	108	106	100	100	106	102	94	100	100	100	102	102	102	94	92	96	102	—
Boston, Mass.	77	72	80	90	82	70	90	80	92	96	88	88	82	76	70	82	78	78	66	74	76	74	74	82	76	72	74	70	66	76	—
New York, N.Y.	70	80	78	86	90	90	90	92	90	92	94	92	88	84	78	76	76	74	70	72	70	68	80	80	76	74	74	72	72	72	—
Philadelphia, Pa.	82	86	88	90	94	96	94	96	96	92	98	96	94	82	88	88	80	80	74	74	80	78	86	86	86	84	88	80	80	80	—
Atlanta, Ga.	92	88	88	90	92	92	90	90	94	92	92	96	94	94	86	88	80	88	74	84	86	78	94	86	84	84	90	84	84	84	—
Washington, D.C.	80	88	86	92	92	98	96	94	98	94	94	96	96	82	86	88	80	80	74	78	82	78	88	86	82	82	78	78	76	80	—
Charleston, S.C.	86	92	96	92	88	88	88	96	94	92	94	88	90	84	86	88	92	90	80	80	84	94	90	92	84	80	86	80	80	80	—
Jacksonville, Fla.	94	84	96	96	92	90	92	92	94	92	90	94	84	94	92	92	88	96	80	86	86	94	78	96	94	94	94	94	82	82	—
Atlantic City, N.J.	74	84	78	78	76	76	74	84	84	96	90	88	84	86	78	80	78	78	74	70	74	74	78	80	76	74	78	74	72	74	—
Tampa, Fla.	88	90	90	92	90	88	90	92	94	94	94	94	94	92	92	90	90	92	90	84	86	82	86	88	90	90	90	86	80	86	—
Mobile, Ala.	80	88	90	92	90	90	92	88	84	94	94	94	94	90	90	90	98	92	94	94	92	90	90	88	86	84	90	86	85	88	—
Vicksburg, Miss.	96	88	92	96	98	96	92	88	94	94	92	98	100	94	98	96	98	96	94	94	98	94	94	90	92	84	84	90	88	92	—
New Orleans, La.	86	88	90	94	96	92	92	88	86	90	94	98	92	90	84	90	92	94	80	94	96	90	84	84	90	92	90	90	88	88	—
Little Rock, Ark.	104	102	102	98	102	104	102	100	96	94	94	98	100	90	98	100	96	92	94	—	100	96	84	84	88	90	84	84	86	90	—
Galveston, Tex.	88	90	88	88	88	88	88	88	88	86	86	88	90	90	90	86	88	88	80	100	—	90	90	88	88	88	94	90	90	90	—
San Antonio, Tex.	—	—	—	—	—	—	—	—	—	—	—	—	—	—	—	—	—	—	—	—	—	—	—	—	—	—	94	88	—	—	—
Memphis, Tenn.	102	96	96	98	100	102	100	98	96	88	96	100	98	98	98	96	98	84	84	90	98	96	88	78	86	90	86	82	84	88	—
Cincinnati, Ohio	88	84	86	98	92	92	88	94	94	88	92	88	86	86	90	82	80	78	72	86	78	88	88	80	82	85	76	76	80	82	—
Pittsburg, Pa.	82	72	84	86	92	94	88	88	94	90	88	80	78	84	86	80	74	74	70	72	74	78	76	78	76	85	72	70	74	80	—
Buffalo, N.Y.	78	72	76	80	80	82	78	80	82	84	78	80	76	80	84	78	70	66	66	70	72	82	72	74	76	78	68	66	70	74	—
Cleveland, Ohio	80	78	82	88	92	86	84	92	92	88	88	82	78	78	86	80	74	70	66	68	70	80	78	72	76	82	70	74	74	78	—
Detroit, Mich.	76	78	86	86	92	90	84	94	92	92	86	86	74	78	86	80	74	72	66	78	70	84	76	76	76	72	70	66	67	78	—
Chicago, Ill.	82	74	86	94	96	90	82	98	92	92	92	88	76	84	82	80	70	74	64	78	74	84	76	72	82	76	68	68	78	82	—
St. Paul, Minn.	84	88	88	100	94	90	92	90	90	88	89	76	82	88	82	72	72	74	76	82	78	72	72	82	80	68	76	84	88	84	—
Des Moines, Iowa	80	86	92	94	92	82	92	90	90	88	86	78	82	88	86	78	78	74	78	82	82	74	84	82	82	76	76	76	76	82	—
St. Louis, Mo.	94	86	92	96	98	98	100	100	98	92	96	94	88	88	96	86	80	78	78	86	96	96	84	86	84	78	80	80	80	88	—
Kansas City, Mo.	88	88	100	96	98	92	102	102	96	98	98	88	88	90	100	94	80	78	74	84	86	90	86	84	86	78	78	80	74	86	—
Omaha, Neb.	81	81	96	96	90	86	94	94	94	92	88	78	82	82	92	78	72	74	74	74	86	80	82	86	80	78	78	84	82	82	—
Los Angeles, Cal.	84	86	80	78	78	78	82	82	82	78	78	82	92	84	82	86	84	86	82	66	78	76	76	76	80	82	82	84	90	90	—
San Diego, Cal.	—	—	—	—	—	—	—	—	—	—	—	—	—	—	—	—	—	—	—	—	—	—	—	—	—	—	—	—	—	—	—

TABLE X.—MINIMUM TEMPERATURE FOR AUGUST, 1896.

Date	1	2	3	4	5	6	7	8	9	10	11	12	13	14	15	16	17	18	19	20	21	22	23	24	25	26	27	28	29	30	31	Nights as hot or hotter than at Phoenix.
Phoenix, Ariz.	76	78	78	70	70	72	78	78	76	80	80	80	82	82	82	68	68	80	76	74	72	76	74	74	74	76	74	74	74	72	73	
Boston, Mass.	57	62	62	66	74	62	70	72	76	76	76	76	72	68	64	64	66	60	56	56	60	64	68	64	64	62	62	54	54	58	56	3
New York, N.Y.	64	66	68	66	70	74	72	68	74	76	76	76	72	70	70	70	64	60	56	56	60	62	64	70	64	62	66	58	58	60	58	4
Philadelphia, Pa.	64	70	66	68	72	74	74	76	78	76	76	78	76	72	72	66	70	62	58	56	60	64	68	74	64	62	66	58	58	60	60	6
Atlanta, Ga.	74	72	74	66	72	72	72	78	74	—	72	76	72	74	70	70	72	72	68	68	72	74	72	72	66	68	66	66	66	64	68	6
Washington, D.C.	60	70	68	70	70	76	76	74	74	74	72	72	74	70	62	70	70	70	68	68	66	76	74	72	66	68	58	66	54	50	60	5
Charleston, S.C.	78	76	80	72	78	76	78	78	78	80	80	76	74	74	74	76	70	78	72	74	70	74	76	78	72	76	74	68	68	64	68	17
Jacksonville, Fla.	72	78	78	76	72	76	76	76	78	78	80	76	78	78	74	74	72	78	78	72	74	74	76	76	78	74	70	72	72	64	64	14
Atlantic City, N.J.	62	68	64	68	68	64	70	74	74	76	74	74	74	76	70	74	74	72	74	78	72	74	70	70	76	74	70	72	64	56	66	2
Tampa, Fla.	76	78	78	78	78	74	74	74	74	74	74	74	74	76	74	74	72	74	74	78	72	74	72	74	76	74	74	72	72	76	72	16
Mobile, Ala.	76	72	76	76	76	76	74	76	76	74	74	76	78	74	74	74	72	74	74	74	78	74	72	76	72	74	68	72	72	76	72	13
Vicksburg, Miss.	78	76	76	76	76	74	72	74	74	76	76	76	78	74	76	74	74	78	78	74	76	76	76	76	76	74	72	70	58	60	72	13
New Orleans, La.	78	76	74	76	76	74	76	76	76	72	76	76	76	74	76	74	76	78	78	80	80	78	78	76	74	74	74	74	74	76	72	17
Little Rock, Ark.	80	80	74	78	76	74	76	76	76	72	72	76	74	74	74	74	76	78	66	62	68	64	64	64	64	68	66	60	62	66	66	11
Galveston, Tex.	80	78	78	82	82	80	80	80	82	80	80	80	80	80	80	78	80	80	78	82	80	84	84	80	78	78	80	78	72	72	76	23
San Antonio, Tex.	74	74	74	74	72	74	72	74	72	76	76	74	72	76	74	74	74	74	74	82	80	76	76	72	70	72	74	76	72	66	76	13
Memphis, Tenn.	78	74	74	78	76	76	68	78	74	70	70	78	78	76	74	76	74	72	62	62	62	70	70	64	68	68	66	60	60	64	68	9
Cincinnati, Ohio	70	68	68	68	70	74	70	72	78	72	74	76	76	70	72	72	60	62	56	56	62	68	66	56	60	62	62	60	58	56	58	6
Pittsburg, Pa.	62	66	66	62	66	74	74	74	74	70	72	72	74	74	66	72	62	56	54	54	58	60	68	60	62	58	52	52	52	52	58	5
Buffalo, N.Y.	58	60	64	68	74	72	70	72	72	72	72	72	64	62	66	72	54	56	54	52	60	60	70	62	56	62	50	50	50	62	54	3
Cleveland, Ohio	64	64	58	62	70	70	72	70	68	68	68	72	66	66	68	70	64	58	50	50	54	60	68	58	56	64	52	54	46	52	60	3
Detroit, Mich.	66	64	64	64	70	72	72	72	70	66	66	66	70	70	68	56	56	54	50	56	58	64	66	60	60	52	52	52	52	60	56	3
Chicago, Ill.	68	64	68	68	68	72	72	72	72	70	68	66	70	66	68	72	64	60	58	58	58	60	58	60	60	64	54	56	60	60	64	3
St. Paul, Minn.	58	60	64	66	70	70	72	72	72	66	66	56	60	64	62	52	52	50	52	54	64	60	50	48	60	44	44	46	60	60	50	1
Des Moines, Iowa	61	62	66	70	72	74	72	72	78	70	60	70	66	64	62	64	58	56	56	56	58	62	52	54	60	52	58	56	58	58	52	2
St. Louis, Mo.	74	70	72	74	78	80	80	82	82	78	76	80	68	74	66	64	66	66	60	64	64	74	60	60	60	68	60	60	62	62	66	2
Kansas City, Mo.	70	66	68	76	76	78	78	78	76	76	76	70	70	68	74	62	58	58	58	66	66	60	60	60	60	58	58	60	60	62	62	9
Omaha, Neb.	64	64	66	78	76	70	76	72	72	66	72	60	70	68	66	62	62	58	56	56	60	64	56	56	56	56	54	58	54	56	56	7
Los Angeles, Cal.	60	60	62	62	60	56	56	54	58	58	62	60	56	58	56	54	62	64	64	62	60	60	60	60	60	58	58	58	56	66	66	2
San Diego, Cal.	62	62	64	66	66	62	62	62	62	64	64	62	62	62	64	64	66	66	66	64	64	64	64	62	62	64	64	64	64	68	66	—

TABLE XI.—TEMPERATURE, 8 A.M., (DRY-BULB THERMOMETER) FOR AUGUST, 1896.

Date	1	2	3	4	5	6	7	8	9	10	11	12	13	14	15	16	17	18	19	20	21	22	23	24	25	26	27	28	29	30	31	Days as hot or hotter than at Phoenix.
Phoenix, Ariz.	80	76	78	74	72	76	78	78	75	78	80	82	82	82	82	74	78	80	74	74	76	76	74	74	78	76	74	77	76	74	73	6
Boston, Mass.	66	64	70	74	78	62	70	70	82	82	84	82	82	70	66	68	70	68	60	58	64	66	70	70	68	62	68	62	60	64	64	5
New York, N. Y.	63	70	70	70	75	74	76	78	82	82	82	76	76	70	72	72	72	68	64	64	64	64	70	72	68	64	68	66	64	66	62	13
Philadelphia, Pa.	68	72	74	76	78	78	80	80	84	84	84	82	82	76	76	74	70	70	64	60	64	68	74	76	68	68	70	66	64	66	64	7
Atlanta, Ga.	78	74	76	76	78	76	74	74	76	76	76	78	80	76	74	76	76	74	72	72	68	74	76	74	68	70	72	72	70	68	70	11
Washington, D. C.	66	72	76	74	78	80	80	80	82	80	76	78	78	74	76	76	72	74	64	60	66	68	72	74	66	70	70	64	62	60	64	23
Charleston, S. C.	82	80	82	78	80	80	80	80	82	82	80	80	78	82	78	76	76	76	68	66	68	78	80	82	80	76	78	76	68	68	72	20
Jacksonville, Fla.	76	82	80	80	78	78	80	80	82	80	78	80	80	80	80	80	80	82	76	78	78	78	80	78	80	80	72	76	76	70	70	10
Atlantic City, N. J.	70	68	72	72	72	72	72	73	76	78	84	82	80	74	76	78	70	70	76	70	70	72	76	74	72	70	74	66	68	66	74	26
Tampa, Fla.	82	82	84	84	84	80	80	80	80	80	76	80	80	78	76	74	78	76	82	78	78	72	76	74	72	70	74	72	80	76	74	20
Mobile, Ala.	78	76	78	80	80	78	80	80	80	78	76	80	80	78	78	78	78	76	82	78	78	80	78	78	84	76	72	82	76	76	74	17
Vicksburg, Miss.	80	80	78	78	80	80	73	76	76	74	74	78	80	78	78	80	80	78	74	76	80	80	78	76	74	76	72	80	76	76	76	22
New Orleans, La.	78	82	78	78	80	78	78	78	80	78	76	78	76	76	80	76	76	80	76	76	76	80	78	80	72	78	78	74	74	74	74	12
Little Rock, Ark.	80	80	78	78	80	80	80	78	78	74	76	78	78	78	78	80	76	66	68	74	74	78	76	64	68	72	66	64	68	68	66	24
Galveston, Tex.	82	82	78	82	84	82	82	82	84	84	78	82	80	82	82	82	84	82	82	82	82	84	84	82	78	82	80	76	72	78	78	13
San Antonio, Tex.	76	76	76	74	74	76	74	76	84	78	76	78	78	78	76	76	74	74	76	76	76	84	84	74	74	74	74	80	72	68	68	14
Memphis, Tenn.	80	76	78	80	80	78	78	80	82	76	76	76	82	72	76	76	70	74	66	66	66	76	78	78	74	74	64	64	64	68	70	5
Cincinnati, Ohio	72	72	74	72	74	74	76	74	78	78	74	74	74	74	78	74	62	58	66	62	68	72	70	60	68	72	58	58	58	58	66	5
Pittsburg, Pa.	66	72	68	68	78	82	78	76	78	78	74	74	70	68	68	68	64	64	58	56	66	62	70	64	62	58	60	54	58	66	62	5
Buffalo, N. Y.	64	64	68	74	74	76	74	74	78	74	74	70	74	68	72	72	58	60	56	56	62	66	72	62	66	66	54	54	58	56	56	3
Cleveland, Ohio	64	63	62	72	76	78	74	76	80	80	72	72	72	74	72	72	60	66	58	56	66	68	72	62	60	56	56	58	60	60	64	2
Detroit, Mich.	68	66	68	68	72	76	80	68	70	68	76	80	68	68	70	72	58	58	54	58	64	68	68	68	64	68	54	60	62	64	64	5
Chicago, Ill.	70	68	70	74	76	80	80	68	80	68	64	72	60	64	62	68	60	58	54	58	64	66	68	62	64	48	54	60	62	60	50	1
St. Paul, Minn.	60	64	70	72	76	76	70	68	72	76	64	60	64	60	62	54	52	52	54	64	68	66	70	66	66	54	52	60	62	60	50	3
Des Moines, Iowa	68	64	70	78	82	84	84	80	72	76	76	78	70	66	60	64	52	54	58	64	76	80	68	62	54	54	52	56	64	60	56	10
St. Louis, Mo.	78	72	74	78	80	72	82	80	78	78	78	82	72	74	76	68	68	68	62	66	80	72	60	62	68	70	52	58	60	66	68	7
Kansas City, Mo.	70	68	70	80	80	72	82	80	72	72	74	70	70	74	74	64	62	62	62	66	72	64	56	60	60	58	58	60	62	62	68	3
Omaha, Neb.	68	66	72	78	76	74	74	73	72	72	74	62	62	66	62	64	64	62	56	66	60	60	56	60	66	58	54	60	66	62	56	—
Los Angeles, Cal.	60	64	64	66	62	58	60	56	56	60	60	60	62	60	62	64	66	66	64	62	60	66	62	58	58	58	62	60	62	74	68	—
San Diego, Cal.	64	64	60	66	62	66	66	60	64	64	66	64	64	64	62	62	66	68	66	66	66	66	64	66	64	62	66	66	68	70	66	—

TABLE XII.—8 A. M., SENSIBLE (OR WET-BULB) TEMPERATURE FOR AUGUST, 1896.

Date	1	2	3	4	5	6	7	8	9	10	11	12	13	14	15	16	17	18	19	20	21	22	23	24	25	26	27	28	29	30	31	Days as hot or hotter than at Phoenix.
Phoenix, Ariz.	70	70	70	70	64	66	70	66	65	66	68	68	66	66	72	72	72	70	72	74	70	66	66	68	68	66	70	72	68	68	66	9
Boston, Mass.	52	64	64	68	72	62	70	68	74	74	74	74	70	66	64	64	62	58	54	52	58	64	64	64	58	60	64	54	54	58	58	12
New York, N. Y.	58	68	64	68	70	72	74	72	74	74	74	74	70	70	68	64	62	58	54	52	58	60	64	64	58	60	64	56	56	58	58	17
Philadelphia, Pa.	58	70	66	70	74	72	76	72	76	74	76	72	72	72	72	70	70	70	64	58	58	66	70	68	58	64	66	56	54	60	58	22
Atlanta, Ga.	72	70	72	72	72	72	74	74	74	72	72	72	72	72	68	72	70	70	64	66	64	63	70	72	68	68	70	66	62	60	58	18
Washington, D. C.	60	70	70	72	72	72	74	74	76	72	74	72	70	70	72	72	70	70	64	66	64	68	70	72	68	68	68	66	62	60	58	27
Charleston, S. C.	80	78	76	74	78	76	74	76	76	76	76	76	76	76	74	76	76	76	76	74	74	76	76	70	64	74	70	74	68	70	70	30
Jacksonville, Fla.	72	78	76	74	74	76	74	76	76	76	74	76	76	76	74	76	76	76	76	74	74	74	76	78	76	76	76	74	76	62	66	17
Atlantic City, N. J.	62	68	66	68	68	68	68	70	68	76	76	76	76	76	76	76	76	76	76	74	74	74	72	72	70	66	68	58	62	58	66	31
Tampa, Fla.	78	78	78	78	76	74	74	76	76	76	74	76	74	74	76	76	76	76	76	78	74	74	78	78	76	74	74	74	74	68	72	30
Mobile, Ala.	72	72	74	78	76	74	74	74	70	76	72	74	74	74	70	74	74	74	74	70	72	72	74	74	72	72	72	64	72	72	70	27
Vicksburg, Miss.	74	76	74	74	76	74	74	74	76	74	76	74	74	74	76	76	74	74	74	76	74	74	76	74	70	72	70	72	72	76	70	30
New Orleans, La.	74	76	74	72	76	74	74	74	76	74	72	72	72	72	72	72	74	74	70	70	72	74	74	74	72	74	72	72	68	72	72	18
Little Rock, Ark.	68	72	70	72	76	68	70	70	70	73	70	72	72	70	72	70	70	72	72	66	72	74	64	64	66	66	56	58	58	60	62	29
Galveston, Tex.	78	78	74	72	76	74	72	74	76	74	74	74	76	76	78	78	76	78	78	78	78	78	78	76	76	76	74	66	74	76	76	25
San Antonio, Tex.	72	72	72	72	70	74	72	72	72	72	72	72	72	70	70	70	72	72	72	74	72	74	70	70	64	56	52	58	62	64	74	18
Memphis, Tenn.	66	68	68	68	70	74	72	72	70	74	72	72	72	70	70	70	68	68	72	66	62	68	68	70	60	64	52	54	56	56	60	12
Cincinnati, Ohio	60	70	60	64	70	70	70	74	74	74	70	70	72	70	70	72	58	58	54	56	58	66	68	58	60	62	52	54	52	56	62	14
Pittsburg, Pa.	56	60	60	64	64	64	74	72	74	76	74	70	71	72	68	72	58	58	54	54	58	66	70	62	64	50	50	54	52	58	62	9
Buffalo, N. Y.	60	66	62	66	64	70	70	72	70	74	70	70	70	70	68	70	56	56	54	54	56	64	68	58	56	62	50	50	50	58	52	8
Cleveland, Ohio	62	62	64	68	70	72	70	72	70	74	74	70	70	66	68	70	56	56	54	52	56	64	68	58	56	60	52	54	50	54	60	9
Detroit, Mich.	68	—	66	68	63	72	70	74	76	74	74	64	70	66	68	70	54	54	50	52	60	66	64	58	60	60	52	54	52	56	60	11
Chicago, Ill.	68	—	66	70	66	62	64	66	70	74	68	64	64	66	68	66	54	54	50	52	56	54	54	58	60	60	54	52	58	58	52	4
St. Paul, Minn.	60	50	66	70	70	66	70	70	66	—	64	56	60	60	58	48	50	50	50	50	58	58	54	50	58	54	46	54	60	56	48	8
Des Moines, Iowa	66	62	64	72	76	70	70	70	70	66	72	56	60	62	70	66	56	60	58	58	66	62	62	58	58	52	50	52	66	58	50	16
St. Louis, Mo.	71	58	70	74	76	74	76	74	74	70	72	58	64	62	70	72	66	60	54	58	72	74	62	58	64	64	56	58	62	62	62	12
Kansas City, Mo.	68	66	66	70	70	68	72	68	70	68	70	72	64	62	72	60	60	58	58	58	72	68	58	56	58	58	56	54	58	60	56	9
Omaha, Neb.	66	64	66	78	68	66	72	68	70	68	64	60	70	60	70	60	56	54	58	64	70	62	52	56	58	50	52	56	64	60	54	—
Los Angeles, Cal.	60	64	62	60	58	56	58	54	54	58	60	54	56	60	60	62	64	64	64	64	60	60	60	54	58	58	62	62	64	68	62	—
San Diego, Cal.	62	62	62	60	60	62	62	58	62	62	64	62	62	64	60	62	64	64	64	64	64	62	64	62	62	62	64	64	66	66	64	—

TABLE XIII.—8 A. M., RELATIVE HUMIDITY FOR AUGUST, 1896.

Date	1	2	3	4	5	6	7	8	9	10	11	12	13	14	15	16	17	18	19	20	21	22	23	24	25	26	27	28	29	30	31	Days more humid than at Phoenix.
*Phoenix, Ariz.	61	74	67	82	65	59	67	53	58	53	54	48	42	42	62	91	75	61	82	100	74	59	66	74	60	59	82	80	66	74	67	19
Boston, Mass.	43	100	92	74	70	59	91	88	68	70	54	68	74	88	86	87	64	52	68	60	65	87	76	76	55	88	84	56	63	72	74	25
New York, N.Y.	54	90	72	90	74	91	95	75	68	55	69	75	83	82	91	82	90	79	89	78	79	100	100	91	90	100	100	68	60	70	69	16
Philadelphia, Pa.	54	88	66	82	83	91	91	66	83	63	69	75	69	82	82	82	40	48	59	58	70	90	82	66	54	81	81	53	61	71	70	23
Atlanta, Ga.	75	82	82	82	83	83	83	75	82	82	73	75	69	82	82	82	74	48	59	73	81	70	91	91	54	100	91	53	64	63	70	23
Washington, D.C.	71	91	91	84	87	77	75	73	73	56	75	65	76	86	87	69	69	61	60	72	78	94	95	85	88	76	90	70	71	81	82	25
Charleston, S.C.	92	92	76	83	92	83	76	84	76	76	84	83	76	76	83	76	83	83	91	67	74	83	83	84	82	91	91	91	77	87	73	28
Jacksonville, Fla.	82	84	83	83	83	82	83	91	91	83	91	83	75	100	91	83	83	83	76	83	83	82	82	91	83	83	91	91	100	72	81	27
Atlantic City, N.J.	64	100	73	82	82	82	90	91	91	84	69	84	75	83	91	82	55	48	76	69	83	82	82	91	84	84	74	61	71	61	73	22
Tampa, Fla.	84	84	83	77	77	83	75	83	91	83	76	83	83	83	75	83	83	76	83	84	83	83	92	92	84	84	76	75	91	82	91	26
Mobile, Ala.	75	82	83	82	77	83	75	75	82	91	83	75	91	83	91	91	83	91	83	100	91	75	91	91	91	82	91	65	91	75	82	28
Vicksburg, Miss.	75	75	83	83	75	75	75	76	82	91	82	75	100	75	91	91	83	91	83	69	68	83	91	91	91	82	91	78	74	74	82	25
New Orleans, La.	83	76	83	83	83	83	83	83	83	83	82	75	83	82	83	82	75	83	83	91	84	83	91	91	91	91	83		74	82	80	26
Little Rock, Ark.	54	68	67	75	61	54	67	76	77	69	82	67	75	67	82	61	74	82	90	90	84	77	74	100	90	73	53	70	79	71	80	18
Galveston, Tex.	84	84	91	75	77	84	84	76	77	69	83	83	83	84	84	76	77	84	84	84	84	83	83	84	83	76	92	59	79	83	72	26
San Antonio, Tex.	82	82	91	82	83	82	82	91	91	82	91	83	83	91	74	74	83	82	82	91	84	91	91	82	82	65	91	82	54	71	72	25
Memphis, Tenn.	68	84	83	82	83	82	82	62	82	82	75	83	83	68	84	76	77	84	80	72	66	75	91	90	80	79	91	54	70	90	72	24
Cincinnati, Ohio	73	90	74	82	83	74	82	91	91	69	83	62	62	91	93	74	72	66	78	78	69	83	91	90	73	79	80	78	78	88	81	22
Pittsburg, Pa.	78	91	83	82	75	82	74	91	93	82	88	87	95	93	93	82	59	70	77	91	61	89	93	91	89	79	70	78	54	90	63	26
Buffalo, N.Y.	60	79	63	66	58	82	91	100	75	100	82	90	87	64	65	82	68	69	77	68	53	70	100	61	50	64	70	60	54	61	77	17
Cleveland, Ohio	79	90	100	73	58	46	82	82	93	91	82	72	81	90	93	91	53	53	59	88	90	90	100	79	78	79	88	88	70	97	79	24
Detroit, Mich.	71	80	82	66	82	82	82	82	73	91	82	72	81	81	65	90	72	70	77	68	61	70	91	61	50	63	77	88	56	78	77	25
Chicago, Ill.	90		73	73	66	68	82	75	83	75	82	91	81	82	90	81	53	59	50	50	64	90	68	71	60	59	55	50	59	70	77	17
St. Paul, Minn.	100	79	81	91	82	71	81	90		81	90	78	79	100	79	64	87	87	76	90	90	90	75	87	64	56	63	88	79	87	87	27
Des Moines, Iowa	90	90	73	75	76	63	91	90		81	82	76	90	82	82	90	68	82	79	90	90	100	88	89	88	88	87	77	79	80	65	24
St. Louis, Mo.	83	82	81	75	76	63	91	69	63	75	83	76	77	82	82	91	90	81	79	70	82	75	71	79	81	72	69	59	77	80	71	26
Kansas City, Mo.	90	90	81	67	61	82	62	54	91	67	67	73	90	82	74	90	89	100	70	90	91	79	88	78	61	69	78	68	89	79	69	23
Omaha, Neb.	90	90	73	100	66	89	89	87	88	82	58	89	81	82	79	79	69	88	73	73	91	90	77	78	61	56	88	78	90	89	88	26
Los Angeles, Cal.	100	90	90	100	71	89	81	88	88	89	89	88	78	100	89	90	90	90	100	100	100	90	100	100	100	100	100	90	72	74	71	27
San Diego, Cal.	90	90	89	71	79	79	80	90	90	80	90	80	81	90	79	100	90	81	81	90	90	80	100	80	90	100	90	90	90	87	90	28

* During August, 1896, the rainfall at Phoenix was 1.77 inches, an excess over the average for seventeen years of .81 inches.

TABLE XIV.—8 P. M., TEMPERATURE (DRY-BULB THERMOMETER) FOR AUGUST, 1896.

Date	1	2	3	4	5	6	7	8	9	10	11	12	13	14	15	16	17	18	19	20	21	22	23	24	25	26	27	28	29	30	31
Phoenix, Ariz.	101	89	79	89	97	99	98	101	105	104	102	104	107	105	108	92	98	101	90	95	99	108	99	100	101	95	86	88	92	108	102
Boston, Mass.	68	64	74	81	64	64	83	76	83	78	74	83	71	67	66	72	70	61	63	65	65	68	72	75	69	68	67	66	63	63	63
New York, N. Y.	68	75	73	83	87	78	87	80	88	85	85	89	84	74	71	77	69	61	67	65	68	68	78	69	74	68	68	69	66	65	63
Philadelphia, Pa.	72	77	76	83	85	87	82	86	89	87	84	90	82	79	78	78	66	66	68	71	73	73	87	72	75	72	81	71	78	79	86
Atlanta, Ga.	78	85	73	84	87	86	86	87	89	90	85	92	82	78	87	83	87	82	81	80	83	75	80	72	74	75	80	78	78	79	86
Washington, D. C.	75	74	80	82	85	87	84	86	90	77	84	88	70	78	80	80	71	70	64	69	75	75	80	63	71	75	69	66	64	70	69
Charleston, S. C.																															
Jacksonville, Fla.	83	87	90	76	82	82	82	83	84	84	82	83	81	83	82	81	83	85	75	77	78	80	79	84	83	72	76	77	74	73	77
Atlantic City, N. J.																															
Tampa, Fla.	84	83	84	84	82	77	80	80	80	86	67	80	80	80	80	82	82	84	84	76	78	80	78	83	84	80	83	79	80	80	77
Mobile, Ala.	77	83	84	80	82	83	80	83	82	83	67	79	85	80	82	83	82	84	87	87	85	81	78	74	81	74	84	79	80	83	83
Vicksburg, Miss.	81	79	77	78	79	79	77	76	76	75	75	78	79	79	76	74	80	77	73	72	81	80	79	76	70	72	72	72	75	77	76
New Orleans, La.	82	80	84	85	90	83	83	76	77	80	84	82	84	82	79	84	88	87	85	92	85	81	79	83	79	81	86	86	82	79	81
Little Rock, Ark.	99	95	93	95	96	97	94	96	76	83	90	94	90	83	95	91	91	80	77	77	94	91	80	80	81	83	80	79	79	83	88
Galveston, Tex.																															
San Antonio, Tex.	94	94	86	87	92	94	85	87	80	90	91	87	87	92	92	78	89	92	93	93	93	93	85	86	92	92	93	90	90	88	88
Memphis, Tenn.	96	88	93	85	92	88	92	96	80	89	91	94	94	84	94	87	86	66	78	75	87	92	72	75	82	74	75	75	76	83	85
Cincinnati, Ohio	70	83	86	78	87	88	92	96	88	89	89	84	84	82	82	79	79	74	75	75	87	92	72	75	77	74	72	70	74	77	71
Pittsburg, Pa.	78	75	79	82	85	82	82	81	72	81	85	78	72	79	80	73	72	74	73	72	74	77	69	73	75	77	68	68	74	77	65
Buffalo, N. Y.	70	70	73	73	77	76	74	80	75	79	76	69	69	76	76	69	66	59	62	62	63	75	72	69	70	61	62	63	66	72	57
Cleveland, Ohio	77	77	83	84	74	76	83	83	74	73	75	77	71	73	76	73	67	67	62	65	66	75	68	69	71	65	68	62	75	75	65
Detroit, Mich.	77	73	77	83	83	74	80	84	77	75	82	71	71	73	71	73	68	64	62	65	68	75	66	70	69	65	62	63	68	71	63
Chicago, Ill.	78	73	81	88	90	74	81	89	89	82	88	73	72	77	79	70	68	64	64	74	63	70	72	68	76	69	66	69	74	72	62
St. Paul, Minn.																					53										
Des Moines, Iowa	77	87	88	88	74	80	90	85	79	85	74	74	76	82	75	70	67	71	70	74	81	67	76	76	76	71	70	72	73	77	71
St. Louis, Mo.	81	83	85	91	93	94	90	93	93	84	92	73	85	80	83	81	74	74	73	81	81	84	72	80	78	73	74	76	75	83	75
Kansas City, Mo.	80	84	94	90	93	88	95	94	92	94	88	82	85	89	94	80	64	71	68	81	90	70	76	77	81	75	71	74	68	82	74
Omaha, Neb.	78	82	91	91	78	83	90	86	89	86	74	76	78	81	73	73	65	73	72	73	80	66	80	80	74	74	73	76	79	79	73
Los Angeles, Cal.	76	75	71	71	69	70	75	75	75	73	72	73	74	75	77	79	75	74	73	71	67	68	70	69	70	70	71	78	85	84	77
San Diego, Cal.																															

TABLE XV.—8 P. M., SENSIBLE (OR WET-BULB) TEMPERATURE FOR AUGUST, 1896.

Date	1	2	3	4	5	6	7	8	9	10	11	12	13	14	15	16	17	18	19	20	21	22	23	24	25	26	27	28	29	30	31	Days as hot or hotter than at Phoenix.
Phoenix, Ariz.	72	72	73	70	72	71	72	72	72	72	72	74	74	74	76	72	74	74	71	74	71	73	75	75	69	67	75	73	73	73	73	5
Boston, Mass.	57	60	68	72	62	63	76	70	77	71	75	75	67	66	64	67	60	56	55	58	58	64	71	64	62	60	64	58	57	56	61	11
New York, N. Y.	64	69	68	73	74	74	73	75	79	70	77	74	72	72	68	74	62	60	58	59	63	68	76	68	66	66	58	59	60	59	62	10
Philadelphia, Pa.	64	65	68	73	77	74	77	75	80	76	76	74	74	73	69	73	58	58	66	56	62	70	71	66	62	63	58	58	58	60	63	11
Atlanta, Ga.	71	76	68	70	73	73	73	76	72	74	72	76	74	72	74	72	72	66	71	71	69	71	73	68	69	74	71	67	67	66	70	12
Washington, D. C.	66	70	70	74	75	76	74	76	80	72	75	77	70	72	71	72	62	60	65	58	65	72	76	62	55	66	56	56	57	61	61	—
Charleston, S. C.																																—
Jacksonville, Fla.	77	78	76	74	78	75	75	74	74	74	73	74	76	75	77	74	73	76	74	73	72	75	76	80	77	71	73	75	64	66	69	24
Atlantic City, N. J.																																—
Tampa, Fla.	76	76	76	78	77	75	76	73	76	77	76	75	76	74	74	77	78	78	79	74	76	78	76	80	78	77	77	76	72	72	73	25
Mobile, Ala.	74	76	78	80	84	76	76	76	75	77	75	73	75	75	74	76	75	77	77	77	78	77	76	78	78	73	77	76	74	75	69	26
Vicksburg, Miss.	74	74	74	74	74	74	72	73	73	71	72	73	73	73	73	71	75	73	69	69	73	75	76	74	69	73	69	56	59	71	71	14
New Orleans, La.	74	76	76	77	76	76	76	70	73	75	75	70	77	72	75	76	78	76	78	76	78	76	75	76	75	74	75	73	74	75	72	25
Little Rock, Ark.	72	74	74	71	72	73	74	73	73	74	72	70	77	72	73	75	75	72	66	73	77	76	72	67	67	69	62	63	74	72	70	13
Galveston, Tex.																																—
San Antonio, Tex.	72	70	73	73	74	72	71	71	72	72	72	72	71	71	73	72	74	74	72	73	73	76	74	74	76	72	70	70	63	65	70	9
Memphis, Tenn.	71	75	78	74	76	76	75	75	78	75	72	72	75	74	78	76	74	65	72	75	75	75	75	63	66	59	63	63	67	65	71	18
Cincinnati, Ohio	68	70	71	72	73	75	75	78	79	76	75	73	75	73	73	70	73	73	57	61	70	76	76	74	66	68	60	60	60	60	59	11
Pittsburg, Pa.	74	74	70	74	75	74	75	77	71	73	72	72	72	72	76	73	60	61	58	61	63	74	67	65	66	70	61	56	62	64	56	13
Buffalo, N. Y.	62	66	67	65	70	62	69	70	70	72	72	63	64	67	70	55	57	54	54	56	59	59	65	59	59	60	55	54	60	65	48	4
Cleveland, Ohio	70	66	70	74	75	71	73	78	82	71	70	63	64	67	71	68	58	60	64	64	71	66	65	66	62	59	59	60	65	57	57	8
Detroit, Mich.	68	65	70	73	75	70	69	80	71	68	77	63	66	67	69	62	57	56	54	57	63	71	59	63	62	58	58	54	61	65	52	6
Chicago, Ill.	70	65	69	76	76	72	76	77	79	77	77	70	70	72	69	63	57	53	53	61	67	69	60	59	64	53	55	60	63	65	57	10
St. Paul, Minn.																																—
Des Moines, Iowa	68	70	78	77	70	74	80	78	70	77	63	64	69	74	74	60	60	60	61	70	76	59	64	64	69	56	60	63	65	64	60	8
St. Louis, Mo.	72	71	73	78	80	78	78	72	79	75	79	64	69	73	76	77	71	72	69	73	83	85	59	66	68	58	59	60	65	68	61	10
Kansas City, Mo.	72	70	77	76	76	76	76	72	73	74	75	74	73	73	76	73	62	64	66	70	79	63	63	65	70	63	61	66	63	68	61	16
Omaha, Neb.	69	72	79	75	70	76	72	80	77	64	68	72	73	71	62	60	60	62	65	70	75	57	65	66	61	59	62	65	71	66	62	10
Los Angeles, Cal.	67	67	64	63	62	64	66	64	64	64	65	61	64	65	68	69	67	66	65	63	62	62	62	63	64	64	65	66	72	71	68	2
San Diego, Cal.																																—

TABLE XVI.—8 P. M., RELATIVE HUMIDITY FOR AUGUST, 1896.

Date	1	2	3	4	5	6	7	8	9	10	11	12	13	14	15	16	17	18	19	20	21	22	23	24	25	26	27	28	29	30	31	Days more humid than at Phoenix.
Phoenix, Ariz.	23	45	79	38	65	24	28	24	19	19	21	22	20	22	33	37	33	28	39	39	24	26	33	32	19	21	60	49	40	26	26	30
Boston, Mass.	51	84	74	62	37	96	74	76	79	70	71	70	70	94	87	87	67	93	60	71	76	100	93	68	68	93	56	52	72	64	89	28
New York, N. Y.	83	73	75	63	56	85	69	78	63	69	70	52	58	91	78	87	67	60	60	71	76	100	70	89	80	93	45	45	57	69	91	26
Philadelphia, Pa.	65	54	66	63	62	63	63	65	68	60	50	47	65	74	78	79	42	61	40	37	66	88	70	89	49	69	61	45	57	50	83	29
Atlanta, Ga.	74	66	79	50	69	50	64	60	68	45	55	48	68	77	56	59	47	42	61	67	49	54	51	82	80	73	54	54	52	50	45	29
Washington, D. C.	64	82	63	67	72	59	63	64	65	36	56	68	100	77	75	59	61	55	59	52	58	54	84	95	80	62	45	52	65	60	63	—
Charleston, S. C.	—	—	—	—	—	—	—	—	—	—	—	—	—	—	—	—	—	—	—	—	—	—	—	—	—	—	—	—	—	—	—	—
Jacksonville, Fla.	76	67	53	91	84	72	72	66	63	59	67	66	80	69	78	71	62	69	95	83	75	78	87	81	76	95	89	91	59	69	67	30
Atlantic City, N. J.	—	—	—	—	—	—	—	—	—	—	—	—	—	—	—	—	—	—	—	—	—	—	—	—	—	—	—	—	—	—	—	—
Tampa, Fla.	71	74	76	77	82	91	81	73	85	67	60	79	80	73	76	82	84	77	80	91	93	88	87	80	78	87	76	87	87	68	80	30
Mobile, Ala.	81	72	77	66	81	76	73	73	74	76	84	82	71	71	81	73	64	70	64	67	73	86	74	91	82	95	56	63	70	65	48	29
Vicksburg, Miss.	63	67	64	50	56	76	73	70	79	76	52	42	60	65	65	59	65	57	52	61	64	55	73	78	81	83	49	45	59	55	47	27
New Orleans, La.	69	85	71	70	53	77	74	75	80	80	58	54	72	78	52	67	64	63	73	87	64	83	83	73	81	72	60	55	56	77	65	28
Little Rock, Ark.	26	37	41	37	30	31	38	33	85	63	44	29	41	60	35	47	47	63	95	83	86	50	68	52	48	49	36	46	39	40	11	24
Galveston, Tex.	—	—	—	—	—	—	—	—	—	—	—	—	—	—	—	—	—	—	—	—	—	—	—	—	—	—	—	—	—	—	—	—
San Antonio, Tex.	34	28	54	52	43	35	50	42	44	41	40	44	45	35	40	75	48	42	37	38	38	46	60	56	34	37	31	37	20	27	40	23
Memphis, Tenn.	36	54	51	61	41	40	51	34	92	52	59	49	41	57	49	67	57	93	60	84	57	45	80	50	55	89	46	52	63	37	50	28
Cincinnati, Ohio	93	71	55	51	51	66	79	66	66	54	52	58	58	58	62	63	43	40	45	46	80	61	67	53	54	72	42	48	43	40	48	27
Pittsburg, Pa.	81	95	64	69	63	79	80	82	72	80	80	87	95	70	82	93	64	57	65	65	72	87	78	62	62	72	65	57	71	58	56	29
Buffalo, N. Y.	64	81	73	66	71	67	78	47	78	77	82	90	73	63	64	93	57	69	35	53	60	47	69	40	52	66	55	55	91	91	48	29
Cleveland, Ohio	67	81	71	66	66	86	79	80	81	82	77	73	79	74	78	94	58	66	69	75	72	82	83	78	75	85	58	84	69	54	61	29
Detroit, Mich.	83	58	71	62	69	82	82	84	74	86	78	78	79	75	93	53	50	60	53	55	80	92	66	70	67	61	59	57	69	72	53	29
Chicago, Ill.	68	65	54	50	53	91	59	58	63	73	60	53	88	94	55	68	49	60	49	47	73	95	49	60	53	32	58	58	51	51	64	28
St. Paul, Minn.	—	—	—	—	—	—	—	—	—	—	—	—	—	—	—	—	—	—	—	—	—	—	—	—	—	—	—	—	—	—	—	—
Des Moines, Iowa	63	58	67	62	82	75	68	73	65	69	52	58	70	68	98	55	85	49	59	80	82	62	53	52	69	37	54	63	65	50	52	29
St. Louis, Mo.	65	57	56	59	57	73	59	63	56	68	57	92	62	62	73	85	85	57	58	71	73	67	44	47	58	38	40	37	65	58	44	27
Kansas City, Mo.	68	50	46	53	48	55	40	44	42	42	62	62	60	59	73	42	73	68	88	81	62	67	40	46	38	38	49	37	73	65	53	28
Omaha, Neb.	63	62	59	47	67	59	65	67	68	56	58	66	72	70	61	57	75	53	69	85	79	58	45	52	48	41	53	55	65	65	53	29
Los Angeles, Cal.	63	66	68	63	67	72	—	55	58	61	69	50	58	58	63	64	66	66	67	64	80	70	68	72	74	72	75	53	53	55	63	30
San Diego, Cal.	—	—	—	—	—	—	—	—	—	—	—	—	—	—	—	—	—	—	—	—	—	—	—	—	—	—	—	—	—	—	—	—

TABLE XVIII.—DAILY OBSERVATIONS TAKEN AT 3 P M., BY THE OBSERVER U. S. WEATHER BUREAU, AT PHOENIX, ARIZONA, FOR THE WINTER MONTHS OF

Day of Month	November, 1896			December, 1896			January, 1897			February, 1897			March, 1897		
	Dry-bulb temp.	Sensi-ble temp.	Rela-tive humid.	Dry-bulb temp.	Sensi-ble temp.	Rela-tive humid.	Dry-bulb temp.	Sensi-ble temp.	Rela-tive humid.	Dry-bulb temp.	Sensi-ble temp.	Rela-tive humid.	Dry-bulb temp.	Sensi-ble temp.	Rela-tive humid.
1	76	56	27	61	46	29	55	44	36	61	51	47	73	51	16
2	72	54	28	64	49	29	54	40	21	65	53	44	68	51	27
3	72	48	10	73	53	22				62	48	32	58	52	68
4	73	52	18	72	50	16	58	40	13	62	48	30	60	48	39
5	71	56	35	70	51	23	61	43	15	60	47	34	61	49	40
6	70	51	24				66	47	17	63	50	38	65	49	29
7				66	48	20	66	48	18						
8	72	54	26	66	48	22	70	50	16	58	47	43	66	51	32
9	75	62	46	74	52	16	52	50	90	57	44	34	64	46	20
10	75	61	44	74	52	16				59	46	34	65	49	26
11	79	62	36	69	50	22	56	55	94	60	45	29	66	48	20
12	81	59	25	70	50	19	66	53	63	58	46	28	62	48	22
13	82	61	26				59	51	57	58	42	23			
14				71	51	20	64	53	46				70	49	16
15	83	59	21	60	48	39	54	49	69	60	44	22	66	49	25
16	80	59	26	60	48	40	54	46	55	66	49	24	65	49	28
17	75	57	31	60	50	48				70	50	19	62	49	42
18	72	53	24	65	50	31	59	46	30	60	50	26	64	50	35
19	74	55	26	68	49	20	57	45	36	59	47	52	61	46	29
20	75	56	26				56	46	40	48		90			
21				73	52	20	65	48	26						
22	76	57	28	75	51	14	71	53	27	58	42	20	54	38	11
23	71	56	35	70	51	23	70	52	26	58	42	19	65	44	8
24	53	52	92	73	54	23				65	49	28	72	50	13
25	56	48	55				63	50	38	76	52	14	77	54	17
26	56	45	40	70	50	19	61	49	41	78	52	11	81	53	9
27	56	45	41				61	49	40				78	50	7
28				68	50	24	63	52	49						
29	57	45	24	52	48	75	60	51	53				61	43	16
30				56	48	57	58	50	55				57	39	10
31				56	45	52							62	43	13

www.ingramcontent.com/pod-product-compliance
Lightning Source LLC
Chambersburg PA
CBHW021958190326
41519CB00010B/1316